Solid Waste and Landfills Management - Recent Advances

Edited by Suhaiza Zailani
and Suriyanarayanan Sarvajayakesavalu

Published in London, United Kingdom

Solid Waste and Landfills Management - Recent Advances
http://dx.doi.org/10.5772/intechopen.100756
Edited by Suhaiza Zailani and Suriyanarayanan Sarvajayakesavalu

Contributors
Terrence Wenga, Abdulkerim Ahmed, Meine Pieter van Dijk, Marcos Negreiros, Eduardo Reis, Augusto Wagner Palhano, Fernando G. Fermoso, Ángeles Trujillo-Reyes, David Jeison, Sofía G. Cuéllar, Antonio Serrano, Soraya Zahedi, Duryodhan Sahu, Manabendra Patra, Muhammad Saleh, Azizan Marzuki, Olawale Theophilus Ogunwumi, Lukumon Salami, Gunamantha Made

First published in London, United Kingdom, 2023 by IntechOpen
IntechOpen is the global imprint of INTECHOPEN LIMITED, registered in England and Wales, registration number: 11086078, 5 Princes Gate Court, London, SW7 2QJ, United Kingdom

British Library Cataloguing-in-Publication Data
A catalogue record for this book is available from the British Library

Additional hard and PDF copies can be obtained from orders@intechopen.com

Solid Waste and Landfills Management - Recent Advances
Edited by Suhaiza Zailani and Suriyanarayanan Sarvajayakesavalu
p. cm.
Print ISBN 978-1-80356-326-8
Online ISBN 978-1-80356-327-5
eBook (PDF) ISBN 978-1-80356-328-2

Meet the editors

Suhaiza Zailani received her MSc in Operational Research and Ph.D. in Management Science from Lancaster University, England, United Kingdom. She is a Professor of Supply Chain at the Faculty of Business and Economics, University Malaya, Kuala Lumpur, Malaysia. Her field of expertise is supply chain management with a focus on sustainability issues, especially in solid waste management. She has taught courses on these subjects for the past 25 years. She has published more than 200 papers in various indexed journals. Dr. Zailani received the Emerald Literati Network Award for Excellence in 2016 and 2019. She was also listed among the world's top 2% of scientists by Stanford University, California, USA, in 2021 and 2022.

Prof. Dr. S. Suriyanarayanan. M.Sc., M.Phil., Ph.D. is the Deputy Director of Research of Vinayaka Mission's Research Foundation (Deemed University), India. Prior to this position, Dr. Suriyanarayanan served as a Postdoctoral fellow at the University of Turin, Italy, as Project Coordinator, Faculty & Head (I/c) at the Department of Water and Health, JSS Academy of Higher Education and Research, Mysuru, India. He also served as Science Officer for SCOPE Beijing Office, RCEES, Beijing, China from April 2015 – March 2017. He is also serving as a Nodal officer for SCOPE India activities under JSSAHER, Mysore. He has research experience in the areas of Environmental Monitoring, Radiation Ecology and Environmental Microbiology. He is a recipient of the Young Scientist research grant award from SERB, DST, Government of India. He secured the prestigious award of visiting scientist fellowship of the Chinese Academy of Sciences during the year 2016-2017. He was recognized as Chairperson for Many national and International conferences. Also serving as Editorial Board Member for peer-reviewed Journals. He has participated in more than 20 national and international conferences and presented papers. He published more than 40 research articles in National and International Journals. He is an Associate Editor of the *Environmental Development Journal* by Elsevier.

Contents

Preface

Management of solid waste refers to the process of collecting, treating, and finally disposing of solid material that has served its function or is no longer useful. This process may also be referred to as trash consolidation. Examples of solid waste materials include wood, paper, plastic, broken furniture, glass, grounded cars, obsolete electronic products, and hospital and market waste. Because most of these waste materials are non-biodegradable, they pile up in landfills where they remain for years. Incorrect disposal of municipal solid waste can lead to unsanitary conditions, which in turn can lead to pollution of the environment and outbreaks of vector-borne diseases (diseases spread by rodents and insects). In addition, unsanitary conditions can make it more likely that people will be exposed to the disease. Landfilling is still the most widespread and cheapest waste management method to dispose of solid waste in the world and is also a major contamination source. Modern landfills are highly engineered containment systems designed to minimize the impact of solid waste on the environment and human health.

Thus, the management of solid waste and landfills presents several technical difficulties to be addressed. Most of the concerns and problems previously associated with solid waste management have been reduced or eliminated because of the improved management of hazardous waste and the emergence of cost-effective, integrated waste management systems with a greater emphasis on waste reduction and recycling. Enhanced air pollution control systems in incinerators have been shown to be effective, and a greater understanding of hazardous compounds contained in solid waste has resulted in environmentally acceptable waste management choices.

Although there have been no revolutionary advances in waste management alternatives, there has been a constant improvement in the technologies required to treat solid waste materials safely and economically. However, they present a myriad of administrative, economic, and social difficulties that need to be managed and resolved. Several studies show that advancements in solid waste management and landfill operations have contributed to lowering the environmental impact of waste disposal and fostering sustainability. This book brings the reader up to date on these possibilities and how solid waste can be managed efficiently and cost-effectively and provides tools to develop and assess alternative solid waste management systems and/or programs. It is organized into two sections: "Waste Management" and "Technologies and Practice." The book includes several images to provide a visual understanding of the various management tactics presented.

Suhaiza Zailani
Professor,
Faculty of Business and Economics,
University of Malaya,
Kuala Lumpur, Malaysia

Suriyanarayanan Sarvajayakesavalu
Deputy Director of Research,
Vinayaka Mission's Research Foundation (Deemed to be University),
Salem, India

Section 1

Waste Management

Chapter 1

Challenges of SWM Hierarchy System: The Stakeholders New Saga in Perspectives

Muhammad Saleh and Azizan Marzuki

Abstract

The book chapter assesses the recent role and advance of private companies/contractors in solid waste management in Kano Metropolis, Nigeria, and the various challenges that ensue as a result of their operations. The methods used include: direct field measurement, focus group discussion, inventory, individual interviews, questionnaire administration, and case study of the operation of private companies in two local government areas of Kano state. The result shows that the total number of 50 inventoried registered companies operating under franchise agreement in Kano metropolis are contributing greatly in terms of efficiency and effectiveness of refuse collection, recovery of material, recycling and disposal, creation of employment opportunities, economic development of Kano Metropolis through real capital investment, among others. The discussion in this paper comes to the conclusion that the Nigeria's solid waste management system requires the adoption of suitable collection, reduction, treatment, recovery, and recycling technologies (solid waste hierarchy). Through a variety of currently accessible scientific treatment approaches, it is necessary to underline the relevant consequences for the potential solutions in MSW at the local and state level. Municipalities (L.G.A.), with the help of the unorganized sector and private waste management companies, are therefore required to concentrate on developing potentials and opportunities in order to achieve the necessary MSWM sustainability for Nigerian cities. This can be done by allowing the stakeholders in SWM to extend their coverage beyond the current operational area to include the entire metropolis.

Keywords: MSWM, challenges, Nigeria, poor economic growth, solid waste private companies

1. Introduction

In recent years, there is an exponential population growth, high density of urban areas, diverse culture, changing food habits, and lifestyles have seen an unresolved problem in terms of Municipal Solid Waste Management (MSWM) in Nigeria, and Kano metropolis was not an exception. As a result, the collection, transportation, treatment, and management of solid waste have become major problems for the municipalities (local government regions). The current work is a thorough analysis that summarizes the current state of SWM, identifies related problems, and derives prospective solutions for MSWM in the Nigerian context. Unsorted solid trash generated at the source by private solid waste management companies, public attitudes, inadequate waste assessment due to an unstructured informal waste industry, unplanned financial, and poor implementation of government legislation are all factors [1]. Also prior to industrial revolution, cities had few materials resources, money was then also scarce and cities and towns had more needs to meet, wastages were also minimized, reuse, recovered, materials were recycled, and organic matters were returned back to soil. During this time, there was an extensive informal recycling and recovery system that flourished, but this began to be displaced by emerging recent and formal municipal waste collection, recycling, recovering, as well as disposal systems, whereby recovery and recycling become large, but almost invisible. As a result of its high rate of generation and poor management, municipal solid waste management is a severe challenge in developing nations. Inefficient solid waste management can lead to a decline in environmental quality as well as the loss of potential resources [2].

However, prior to the formal involvement of private corporations in the waste management system in Kano metropolis, which dates to the 1990s, a small number of businesses functioned informally under open competition with varied degrees of success. As a result, their effects were felt more by the individual customers than by the entire refuse management system. In order to address this worrying situation, the state government in the middle of 2000 hired a single private company to handle solid waste collection and disposal in Kano Metropolis for a fee of roughly 200 million Naira (=N=200,000,000.00), or roughly $1.17 million (US), annually. The company's contract with the government was terminated in less than 2 years as a result of its own inability to produce the required and anticipated results. This study reviews current advances in solid waste management hierarchy challenges in Kano metropolis-Nigeria and the need for an effective collection, recovery, and recycling policy. This was carried out through the use of direct field measurement, focus group discussion (FGD), inventory, individual interviews, questionnaire administration, and case study, and the use of large availability of secondary data. The challenges were mainly the government inability to deal with the waste due to growing population, and lost in state revenue, as such the need for the government to diversify by allowing private stakeholders to join the wagon. Our review is based on 85% literature searches and personal field surveys. Observations revealed that the findings were caused by the careless way in which stakeholders and Nigerians alike managed solid trash, as seen by the recent collection rates of less than 40% of the entire amount of solid garbage generated (i.e., approximately 30% was collected). Therefore, the country's large amount of solid waste could present potential for material recovery, reuse, recycling, and reduction, which would enhance the national economy and bring Nigeria closer to sustainable waste management, with an efficient collection system and proper policies. Municipalities face difficulties as a result of the inhabitants' lack of awareness of the problems with trash management and their careless behavior with their waste. The hazard posed by MSW at landfill sites, which emits dangerous greenhouse gasses and subsequently pollutes the environment and contaminates groundwater through the creation of leachates, must be addressed.

The production of dangerous chemical wastes by cities, such as hospitals and factories, which now causes breathing issues and early deaths, is another recent development in MSWM [3]. Nigeria has recently emerged as a market for recycling, despite the fact that recycling has not met the required standards [4]. Poor MSW management at landfill sites frequently draws animals, rodents, mosquitoes, vultures, and scavengers, which could have a negative impact on front-line workers' health or even result in their demise [1]. The development of technical knowledge on the conversion of waste to resources would be facilitated by education and awareness programs, which are crucial in this situation. This would educate more people to adopt new technology strategy skills in areas in the management of waste, while embracing all of these strategies' and advantages 2014 (Ibrahim). Furthermore, the private stakeholders, from both formal and informal enterprises of widely varying sizes and capabilities, can supplement the knowledge and capacity of the local authority to implement recycling, manage organic waste and serve households with waste reduction, collection, recovering, and recycling through private solid waste management contribution, which is the main objective of this book chapter.

Finally, a few cities in Nigeria such as Kaduna, Ibadan, Abuja, and Lagos have showcased the positive intent toward MSWM strategies as per solid waste hierarchy selection, which have been discussed in this study. It has also been observed that the municipalities are focusing mainly on the collection part, while ignoring recent advance treatment methods. However, this also needs further upgrade in order to eliminate the MSWM issue [1]. The resourceful material recovery and other processes alike have been a challenge that could be achieved by the help of the informal sector in the MSWM process, which is fully discussed in this study. However, this idea can only be fulfilled with support and funding from the government agencies, public awareness, participation, and to eliminate the citizen attitudinal problem. The mention of private solid waste management companies under one umbrella could contribute in a step toward clean sustainable cities. Hence, the present study is a comprehensive review carried out in all the possible strategies from past to future pertaining to MSWM by private companies, government agencies, and other stakeholders alike, also the study addresses the challenges and potential opportunities for the future urban cities of Nigeria.

2. Challenges of recovery, recycling, collection, and disposal of solid wastes by private company

2.1 Description of the practice

2.1.1 What the practice

This practice is the result of recent challenges in the recovery, recycling, collection, and disposal of solid wastes by private individuals and businesses. In this instance, the activity is being carried out by a registered solid waste private company in Nigeria's historic city of Kano. It is a limited liability company undertaking collection and disposal of solid wastes on a commercial basis. Its services include on site separation of these waste before transporting the collected waste via a temporary site before final disposal, this company also undertakes in the sweeping of street/market, supply of refuse bins, collection and disposal of refuse. At present its services are available to individuals, private, and governmental organizations within the Kano Metropolitan area. The customers of this company are typically home or company owners who currently have a contract with the refuse collection and disposal company that specifies the anticipated volume of waste to be collected on a regular basis and on the basis of which the company informs the customer of the monthly fee. Currently, a client will pay N2, 000.00 a month on average to have a single 200-liter drum utilized as a waste can and cleaned out once a week. The refuse is being disposed of through recovery, recycling, and leveling of what is left into a borrow pits in designated areas approved by the Kano State Urban Planning and Development Authority (KNUPDA). They operate a workshop for the maintenance of their machines and equipment. Kano is a developing city in Northern Nigeria that is expected to become economic vibrant from the activity of this stakeholder activity in the coming years, this of course is regarded as a recent advance for a developing nation in this twenty-first century era.

2.1.2 The purpose of the practice

To support the efforts of the local and state governments, the Ministry of Environment, and its agency KNUPDA in keeping Kano clean, the aforementioned practice, which involved a private enterprise that is purely engaged in an efficient and scientific SWM, was formed. The creation of waste disposal companies was prompted by their realization that they might profit from the collection, recovery, recycling, and disposal of wastes. The different stakeholders include the waste disposal company and their customers—owners of residential houses, hotels, markets, hospitals, private and public institutions, industries, companies, the local authorities, and Ministry of environment with the new agency Cape gate a privately owned German company hired by the state government in June 2021 to serve as an overseer but to be answerable to the state ministry of environment, therefore purpose and practice are hereby state as in:

2.1.3 Research question and objective

The research question is:

- What is the importance of private sector involvement in MSWM?
- How can the government and other SWM stakeholders control SWM practice in the study area?

The research objective is:

- To investigate the importance of private sector involvement in MSWM practice
- To learn how government agencies and SW stakeholders influence SWM practice in Kano metropolis

The choices made in the process of execution of the study can be justification in essence that the study is one of the many attempts that have been made by authors in order to resolve the SWM related issues, but still, solid waste management is not being handled holistically, this is because most of the cities in Nigeria are still undertaking collection of mixed type of waste [5], while the enforcement of appropriate local/state government treatment system is found missing. Eventually, there are recent advanced issues of MSWM in Nigeria that have not been addressed efficiently so far for the treatment of waste. Hence, this paper also discusses the missing interlinks and loopholes in this regard. In fact, the situation has become severe for most of the municipal authorities as a proper assessment of the SWM is not done before suggesting and implementing the strategies [6]. Therefore, the end of resourceful solid waste leads to unscientific dumpsites without proper treatment of waste. However, a detailed and well-justified table of the important points from the large and available literature was also investigated, and the following aspects ensure (**Table 1**):

Stakeholder	What to do
Government	The government's primary emphasis on SWM is minimizing and eliminating the detrimental impacts of waste materials on the environment and human health in order to foster economic growth and a higher standard of living. This needs to be done as soon as possible to reduce cost and avoid waste accumulation.
Private sectors	To get in the proper solid-waste collection, recovery, recycling and reducing of waste is important for the protection of public health, safety, and environmental quality. It is a labor-intensive activity, accounting for approximately three-quarters of the total cost of solid-waste management.
Private companies	They are to collect waste as well as encourage recycling in this manner: • Keep recycling bins close by. • Get recyclable items in supermarkets. • Reduce the price of plastic bags and create reusable ones. • Make it fashionable to recycle.
Citizens/ community leaders	Get involved in solid waste management in your neighborhood. This can be achieved through public participation, awareness campaigns, town hall meetings, incentives for good SWM and so forth, while community leaders should serve as watchdogs as well as those that ensure a clean and aesthetic environment.
Visitors	Visitors to a certain neighborhood most also strictly observes all the do and don'ts of the community they found themselves
Informal sectors	This includes all stakeholders such as scavengers, waste pickers, itinerant waste managers, etc. these stakeholders have time without numbers being acknowledged to be important in SWM of a city. The unorganized informal sector of waste needs to be integrated with the formalized system. Also, MSW is generally handled by untrained staff without safety equipment, which needs to be uplifted immediately.

Source: Authors initiation 2022.

Table 1
Justification for SWM stakeholders in solid waste management perspective.

2.1.4 Who initiates the recent practice?

The owner of the private liability company claimed that he initiated the practice in two other cities in Nigeria and later opened the company in Kano. In fact, most of the waste disposal companies registered in the study area were initiated by their owners. They first began to enlighten the communities on the need to have clean environments in the course of which they were able to have more customers. For example, one of the companies, a private solid waste enterprise, was given a contract by Kano State government for the collection and disposal of wastes in Kano City and the state government through the ministry of environment issued a circular to the residential houses, informing them of the engagement of the company and on the need for the citizens to patronize it. And in complementing the effort of the state government, these operators also embark on awareness and enlightenment campaigns by advertisement in radios and televisions. This has resulted in having more customers than the earlier envisaged. A recent advance in this regard is that these registered private solid waste companies send questionnaires to their customers to fill, with the sole aim of seeing that they serve their customers better. And once in a while, they organize public meetings with their customers to air their views on aspects of improvement or otherwise.

2.1.5 The management of the waste private company

It was discovered that the managing director of the other registered private waste firm and the new SW German private company Cape Gate Company (overseer) may also be the owner of these enterprises. In order to promote economic growth and a higher standard of living, the government places a key priority on SWM: minimizing and eliminating the negative effects of waste materials on the human health and environment. In order to save money and prevent waste accumulation, this needs to be done as soon as feasible. As far as operation and maintenance are concerned, the enterprises currently self-finance using bank loans. On-site engineers and sanitation inspectors who have received the appropriate training typically handle waste material recovery and recycling.

2.1.6 The beneficiaries

Due to the issue of secrecy, as such it is very difficult to find out the number of people served by the operators of the private waste collection and disposal companies, this issue is so rampant among all the waste companies visited by this author. However, the local government areas mostly covered by these operators are Kano Municipal and Nassarawa local government areas. The primary beneficiaries are the end users (waste generators) and the private operators the (SW managers). And this is followed closely by the local and state governments who are also beneficiaries in that these private companies' complements their efforts of achieving a clean environment in Kano metropolis and as well as monetary aspect these private companies must pay to the state government every year as a public liability company.

2.1.7 How long has the practice been in existence?

This practice started as far back as 1984, but the recovery and recycling aspect started less than a year ago when the then Military Government launched a program titled "war against indiscipline." Lack of environmental sanitation was among the first items attracting severe penalties that included: heavy fines, twelve (12) strokes of cane, imprisonment or both. In fact, most household owners have welcomed the coming of these companies, especially the overseer company that assisted in complying with the rules and regulations introduced under the program from the initial kick off.

2.1.8 Where it is being used and its spread

Apart from the German company hired by the Kano state ministry of environment. These other solid waste private operators or waste collection and disposal in Kano walled City. There are a few popular ones that are also situated in Kano. These include Rubbish and Garbage Limited, Foundation and Waste Disposal Company (WASCO), and Clean Town Engineering Company (Nigeria) Limited. Most Nigerian cities, including Lagos and Kano, have this technique in place, but it's unclear whether it was introduced there because Kano City's operators claimed it was their own idea [7]. One of the companies, WASCO, however, began its operations in Lagos before relocating to Kano in April 1998. There are few common ones also located in Kano and this include: Clean Town Engineering Company (Nig.) Limited, Foundation and Waste Disposal Company (WASCO) and Rubbish and Garbage Limited. This practice has been operational in most Nigerian cities like Lagos and Kano, but it is not clear whether the system was transferred from these cities as the operators in Kano City said it was their personal initiative [7]. However, one of the companies, known as WASCO was initially operating in Lagos before coming to Kano in April 1998. The system is spreading throughout Kano City on a daily basis, but due to the failure of government agencies REMASAB, a new private German company will be taking over the management of these organizations, which will require the companies to gain beyond its current status of material recovery and recycling from municipal solid waste in Kano metropolis.

2.1.8.1 Tools and methods used

The private operators companies even though they have limited areas of coverage, these enterprises have been found to have been very effective in their operations. The private operators and the government agencies such as KNUPDA, REMASAB, and Ministry of Environment and of Cape Gate were all found to be involved in public education/awareness campaigns and enlightenment on the need for recovery as well as the importance of recycling recovered waste. They inform and enlighten the general public as well as individuals, families, and or community solid waste stakeholders/groups to improve their environmental health and support the projects of the state government in the area of clean environment and the monthly payment of the agreed fees to the private company. Whereas the benefits of these public education and enlightenment campaign have been recent advances fully highlighted as shown below:-

- Environmental education to create awareness on the consequences of poor hygiene and sanitary practices.

- Community/neighborhood sensitization and education on privatization, partnership, and public.

- Participation of the whole public in general sanitation schemes.

- Developing and sustaining individual responsibility in paying sanitation charges, protecting and maintaining public facilities wherever they might have been located.

The benefits of public knowledge are manifold. To effectively reap the benefits, sanitation authorities should develop strategies and procedures for putting them into action. One of these is identifying particular target audiences and planning the methods and media to reach them. Second, should environmental education be incorporated into the curriculum or should it only be used in extracurricular activities like theater and debate? How do we acquire the tools and funds necessary to support the campaigns? Will business stakeholders and charitable people be willing to sponsor the campaigns, or at the very least some radio and television jingles? The benefits of public enlightenment are so many. However, to effectively reap them, sanitation authorities should identify strategies and work out means of achieving them. One of such is the identification of specific target groups and preparing the means and medium of reaching them. Secondly, can we include environmental education in school programs or limit it to extracurricular activities such as drama and debating? How do we find material and financial resources to fund the campaigns? Will corporate stakeholders and philanthropic individuals be willing to sponsor the campaigns or at least some jingles on radio and television? Or limit it to extracurricular activities such as drama and debating? How do we find material and financial resources to fund the campaigns? Will corporate and philanthropic individuals be willing to sponsor the campaigns or even if it is some jingles on radio and television? [8]. Most especially in the newly introduced on-site recovery and recycling of the recyclables from the waste stream, which has been a challenge for sometimes in the study area until of recent. Therefore, if the solid waste management hierarchy tool concept is fully utilized, the rest is simple; this is because waste will have been reduced by almost 35–40% [9, 10]. A descriptive cross-sectional study was conducted from June to August 2021 in two of the local governments in Nigeria. Kano, Nassarawa, and Municipal local governments are among the 774 local government areas in Nigeria and these two Local Government Areas are within the Kano urban Area in Kano State, Nigeria. Its headquarters are in Kofar Kudu and Giginyu, in the south part of the city of Kano. It has an area of 17 km^2 and 18.8 km^2 and a population of 465,525 and 464,321, respectively, as at the 2011 population census. The main environmental sanitation scope given attention in Kano Municipal and Nassarawa local government areas is waste collection and disposal. Kano Municipal and Nassarawa, as urban centers, are known for their overpopulation and commercial activities. Kano Municipal and Nassarawa like many other cities in Nigeria are faced with perennial ecological problems favorable to the survival of parasites causing diseases. The major sanitation facilities include: REMASAB bins, private waste management companies, and other SWM stakeholders engaged with evacuating tanks, personal bins, basket, unauthorized dump site plot, and occasionally, fumigation of the markets. The study contained three domains mainly; stakeholders and vital information, mode of refuse disposal, and refuse disposal consistency. Waste that was observed in the study area is mostly from residential areas. Mode of refuse disposal is focused on collecting information on methods of refuse disposal, types of building residents live, and material use for disposal. While refuse are collected and disposed of at a certain fees collected every month by the private companies. Study questionnaires were also used to collect data from the study area. The data sources were measured using a structure questionnaire. Primarily, a structured interview questionnaire, which was made valid and reliable by a team of other solid waste management experts, was used for data collection. The final questionnaire had 16 questions covering information about domains; information and vital information, mode of refuse disposal, and refuse disposal consistency.

2.1.8.2 How the practice is being implemented

The customers who are the citizens are expected to always clean their houses and its surroundings and their establishments and then put the solid wastes in the supplied drums form the private companies. Items to recover from the waste are the usual recyclables of metal and aluminum. Glass, paper, (non-degradable) and the kitchen / food waste which constitute the highest in the waste stream (degradable) [11]. They are to also to ensure that payment for services rendered is regularly made and on time. The waste collection and disposal companies on their part are to regularly convey the solid wastes from their customers' residences and business premises for proper on-site recovery and the recycling process and finally to dumping sites. They are also to enlighten the general public on keeping a clean environment. As for the cost per household, for the service rendered, the waste disposal companies' charges are dependent on the distance from the houses to the disposal sites. At the moment it is a flat rate of N2, 000 per month based on their limited area of coverage for example the Clean Town Company initially started operating using 200 liters drum,

However, a 100 liter drum was provided for recyclables with the current advancement in recovery and recycling; these drum garbage bins are typically located in its client's compound. Using pickup trucks, the firm collects the filled drum bins at predetermined intervals and replaces them with clean ones while transporting the filled drum bins to one of the KNUPDA-approved dumping grounds for cleaning [12]. But with this recent advance of recovery and recycling a 100 liters drum was provide for the recyclables, these drum refuse bins are usually placed in its client's compound. At an agreed interval, the company collects the filled up drum bins using pickup vehicles and replaces them with clean ones, while the collected filled up drum bins are transported and cleaned at one of the KNUPDA approved disposal grounds [12]. However, as the demand for its services increased, that is, more residencies and business premises joined, it became uneconomical for the company to be moving with replacement bins. Therefore, these private companies decided to change it mode of operation by using tipping vehicles such that the contents of every drum refuse bin is emptied into the tipper once and for all and the cleaned bin returned to the client's compound instantly. At the end of every month the company's revenue collector goes round to the clients to collect the agreed fees and issue receipts [8, 13]. The solid waste private operators of this new system of waste tipping are in charge of the operation and maintenance of their equipment and machines after the day's operation. There are many reasons why this practice is set up this way; these include:

a. REMASAB being the government agencies that are charged with the responsibility of solid waste collection and disposal have not succeeded.

b. The customers have been demanding private sector participation because of their effectiveness has it was seen in the low density residential area of Kano metropolis the GRA.

c. The local and state governments have funding limitations, where only 5% of the state budget is allotted

d. Present government policy also encourages private sector participation in wastes collection and disposal although, it was found out that apart from the muted secrecy regarding number of customers, and individual companies are also at loggerhead with one another.

3. Analysis

3.1 The success of the practice

Depending on the viewpoint of the stakeholder Saga, the new and recently implemented on-site recovery and recycling of recyclables practice as well as collection and disposal of solid waste by individual or private operators/companies can be regarded to be successful or not. In general, where these private individuals or private waste companies operate (i.e., these are waste stakeholders), their places where they are engaged solid waste disposal are generally clean drastically, thereby reducing epidemic in these areas, and this was the reason why some citizens prefer their service to those of the government. The Kano State government, the Kano State Urban Planning and Development Authority, the eight local governments in the Kano metropolis, the

Ministry of Health–Kano State, private companies, customers of solid wastes disposal outfits, individuals or companies, people living in peri-urban areas, industries, educational institutions, and public/semipublic land uses are among the stakeholders who are actually impacted by this practice. The analysis's findings and analyses are completely stressed in the following manner:

3.1.1 Result and discussion

Kano State Government/Kano State urban planning and development Authority Information was sourced and finally collected from KNUPDA Headquarters through a personal visit of the author. Therefore, the Kano State government through its agency KNUPDA/ministry of environment and since part of its functions is to keep Kano city/eight other LGA within the metropolis clean, through the new introduced SW recovery and as well recycling of recyclables and collection and disposal of these solid wastes. However, because of limitations in funding, REMASAB cannot cope [14]. The agency sees the operation of the private individuals or companies as complimentary. Therefore, in **Table 2**, REMASAB as a government agency that was formerly in charge of SWM has this to say to private operations or private companies dealing in solid waste management in the study area (**Table 3**).

3.1.2 Local government area

In all the eight local governments that make up Kano metropolis Municipal, Dala and Gwale Ungogo, Tarauni, Kumbotso, Nassarawa Dawakin Topa, and kudu local government area that make up the study area. Only Nassarawa and Municipal local governments, where the private operators are active, were used to filter and collect the information. According to the constitution, local governments should be in control

Assessment	REMASAB management	REMASAB administrative	REMASAB supervisors	REMASAB field workers	Total %
Efficiency	30	15	22	33	100
Effectiveness	36	30	12	22	100
Reliability	12	18	3	10	43
Accessibility	20	10	20	17	67
Total	98	63	67	82	75.5

Source: Authors field survey 2021.
Efficiency – in this aspect private waste companies/contractors are efficient in most areas where they operate but are costly, and as such the poor citizen may find it not affordable to patronize them. This is because their enlightenment programs are limited in scope, however, all sample staffs under REMASAB believed (100%) that private waste company are very efficient more especially in their of operation, but in terms of reliability (43%), this is very bad, due to the fact that the assessment depends on monthly collection from customers patronizing these companies. Effectiveness – in this aspect private waste companies are again effective, since they are operating based on the objective of having a clean environment that must be achieved in their areas of operation. Even though they have, however, a limited number of equipment and vehicles (infrastructure) especially waste disposal vans and trucks. Reliability – in this aspect the table above shows that they are not very reliable, as any default in the monthly payment by the customers will result in the stopping of the services to such customers. The accessibility aspect is they are accessible as much as these customers are ready to pay for services rendered with regard to SWM, this is the reason the assessment in the table above is 67%, such agreements are said to be recent advances in SWM in the study area.

Table 2.
An assessment of solid waste companies by REMASAB.

Assessment	Nassarawa local government area	Municipal local government area	Gwale local government area	Tarauni local government area	Total %
Efficiency	21	18	21	10	70
Effectiveness	20	10	30	15	75
Reliability	15	20	10	10	55
Accessibility	24	12	20	10	66
Total	81	60	81	45	72

Source: Authors field survey 2021.

Table 3
An assessment of solid waste companies by local government area.

of SWM throughout their jurisdictions. However, because of urbanization and population growth, these local governments now view the operations of the private operators as complementary. As a result, these local governments have the following to say about the operations of the commercial garbage operators in these two local government areas.: Efficiency aspect even though they are efficient in solid wastes recover, recycling, collection and disposal but find it very difficult (61%) to expend their profits on health, sanitation and public enlightenment. While in the aspect of effectiveness these private waste companies are effective among the individual houses, companies, hospitals, hotels and markets where they operate. Although, they are not ready to operate in peri-urban areas (48%) except with government support. Reliability aspects: these private SW companies are reliable but their area of coverage is small and may need government support for expansion (64%). In the area of accessibility these companies will be needing stronger trucks to operate in peri-urban areas because of bad roads and drainage, this will be accompanied with fees reduction in such areas (**Figure 1**).

Figure 1.
Nature of waste in the two (2) LGA OF Kano state Nigeria.

3.1.3 Ministry of Health, Kano State

This ministry is responsible for public enlightenment on health and sanitation matters and this includes solid waste in the state. Information was obtained from the Ministry of Health headquarters in Kano through my personal visits to the ministry headquarter located at Audu Bako secretariat. The ministry has this to say of the waste private operators (**Tables 4** and 5):

Assessment	Staff in the ministry	Field supervisor	State health inspectors	Hospital waste managers	Total %
Efficiency	30	18	16	12	61
Effectiveness	18	12	14	20	48
Reliability	11	19	21	13	64
Accessibility	28	10	15	16	64
Total	87	59	67	61	68.5

Source: Authors field survey 2021.
Efficiency – They are efficient in those places where they are operating. However, despite ambitious coverage growth, they lack the necessary equipment for managing hospital waste (12%), and these enterprises are ineffective at educating the public about health and sanitation (18 percent). The only people that promote their services are their clients. Due to their inadequate equipment, they cannot be relied on to serve the entirety of Kano City in their operation (21 percent). Accessibility (15) Residents of Kano City's pre-urban (unplanned) neighborhoods cannot access them. The primary causes of this are inadequate access roads and drainage systems.

Table 4.
An assessment of solid waste companies by Ministry of Health, Kano State.

Assessment	Nassarawa neighborhood I	Municipal by Taludu junction	Nassarawa neighborhood II	Municipal by K/ Kabuga junction	Total %
Efficiency	44	21	0	0	61
Effectiveness	29	19	0	0	48
Reliability	48	16	0	0	64
Accessibility	50	14	0	0	64
Total	98	63	67	82	75.5

Source: Authors field survey 2021.

Table 5.
An assessment of solid waste companies operation by customers of solid waste companies, Kano State.

3.1.4 Customers of private operators

Customers of private operators include market vendors, hotels, hospitals, markets, and factories, among others. While the clients are paying for the services provided, the private operators collect and dispose of the waste from the customers' locations. Regarding the private operators, they have this to say. Although pricey, they are effective. Effectiveness. They arrive to collect and dispose of the rubbish as soon as and when required. They are dependable since they always make an effort to fulfill their own obligations under the contract. Accessibility. These private businesses are open to everyone with the means to pay for the necessary services (**Table 6**).

Assessment	Nassarawa Tudun Murtala	Municipal by Emir Palace	Bakin Bulo neighborhood	Municipal by Kofar Mazugal	Total %
Efficiency	23	10	10	6	49
Effectiveness	31	20	10	15	75
Reliability	25	11	10	12	53
Accessibility	10	12	10	13	45
Total	89	53	40	46	55.5

Source: Authors field survey 2021.

Table 6.
An assessment of solid waste companies by people leaving in peri-urban areas of Kano.

3.1.5 People living in peri-urban areas of Kano walled City

These are the unplanned areas where the low-income earners live. Information was collected from these people through oral interviews. They have this to say of the private operators as formal solid waste stakeholders that operate within these two earlier mentioned local government areas. Efficiency: Given their high cost (49%) and inability to be accessed by those with low incomes, they cannot be claimed to be efficient. Effectiveness because these locations are continuously maintained clean, they are effective where they operate (75%). They have a good track record when it comes to their coverage (53%). Considering that the majority of them (45%) living outside the two local government areas where the private garbage operators are not allowed to operate, low-income earners do not have access to the private operators at any time they desire to. Based on this analysis and discussion, the result of study of these forms of waste stakeholders operation in our cities in recent times has added to some of the recent advances in solid waste management in developing countries in this 21st century of our times.

3.2 Impacts of the practice on the study area

Efficiency: Given their high cost (49%) and inability to be accessed by those with low incomes, they cannot be claimed to be efficient. Effectiveness because these locations are continuously maintained clean, they are effective where they operate (75 percent). They have a good track record when it comes to their coverage (53 percent). Considering that the majority of them (45%) living outside the two local government areas where the private garbage operators are allowed to operate, low-income earners do not have access to the private operators at any time they desire to. The participation of the private operators have helped in reducing the heaps of refuse in some locations in the city, this is an expected outcome. Although, the practice has no impact on low-income areas of the city. And the reason behind this was found to be associated with the fact that the low-income communities are either unwilling or unable to pay for the services rendered by these private operators. Most of the households contacted on their willingness to pay for commercial refuse collection and disposal claim that if they had money, their priority would be for payment of water and electricity and not waste, even though these two facilities are in a very short supply to these two local government areas. However, with vigorous public education about the risks of having an unhygienic environment owing to improper management of solid waste (SW), as advocated by the current chairman of the Nassarawa Local Government Council, people will be more prepared to pay for solid waste collection and disposal. By building access roads to these peri-urban centers (unplanned areas), the government can also encourage the private operators to begin operations there.

When asked if they had any plans to operate in such places in the near future, the managing director of the new company Cape cate managing garbage in Kano metropolis responded that they had already begun discussing the procedures with various Local Government Authorities in these urban areas. He did, however, offer the Sabon Gari market as an illustration of his company's presence there. With the assistance of the market authority, they now have a contract for the sweeping and disposal of market waste, and they are paid from the levies that are collected from individual traders. The managing director claims that this arrangement has already achieved over 80% success and is now self-sustaining.

On-site source recovery and recycling have been found to provide income for the owners and provide employment for locals. These autonomous companies work clandestinely. Out of anxiety that more private companies might enter the waste collection and disposal market, they are reluctant to freely provide information about how they conduct their operations. The cost of waste collection and disposal will substantially decrease with increased public awareness and private sector involvement, making it possible for people with modest incomes to afford the services of private managers/operators.

3.3 Assessment of private operators' activity

The activity of the private operator in Kano city came into limelight some decades back, precisely around 1983, during this period, the private contractor stars with only 50 houses because not all residential could afford it, but, after some period of public enlightenment , more houses decided to also joined the wagon mainly because they have seen the impact of these private operators on their surroundings, at the moment, based on personal counting of the various 200litre drum used as bin, the first private contractor (Clean Town Engineering Company) has more than 2000 of the 4500 drums counted In other words its area of operation in terms of coverage has increased. As a result of the success of this practice, more private operators have entered into this business. Whereas, the practice of solid waste collection and disposal by private operators is also operational in some parts of Lagos, Kaduna, Port Harcourt and Onitsha cities of Nigeria. Presently, there are more than 10 out of the 52 registered waste contractors operating in these two local government areas. These 10 private companies have all taken the bull by the horn by including proper recovery of waste materials from the waste stream before final disposal at the designated landfill located within Kano metropolis. It should be noted that these private operators have both medium and high-income earners as their customers. However, they do not operate within the low-income areas because people in those areas cannot afford their charges. For this stakeholders practice to become more effective in the low-income areas, then a bigger stakeholder like the state and local governments need to also embark on more public enlightenment to this residents on the need for them to keep their environment clean, and free from germs, as such, the government as part of its responsibility to its citizens need to dialogue with the private operators to subsidize their rate so that collection and disposal of solid waste by the private operators in the peri-urban centers can also be radicalized for a productivity as well as replicability [15]. This can be considered recent advances in SWM, if fully implemented in these two local government areas.

3.4 Assessment of sustainability of operation

Presently, the private operators are operating independently without assistance or cooperation from the government or communities. They are registered companies that operate like any other profit making Liability Company because late last year some of the private companies started selling shares to its customers in their area of operation. In other words they have their institutional arrangements with their customers for the total management, operation and maintenance. Thereafter, profit will be shared among those that subscribe into the business, although they do not have support arrangements or external influences or inputs from outside the community they served. The practice as of now can only collapse when the commitment of the owners (owner and customers) is no more there or the government commits enough finance to its agency responsible for solid waste collection and disposal. And from the look of things these conditions are not realistic considering the enormous responsibility the government shoulder on a daily basis.

With my several personal visits to the operation areas as well as the company offices, I can out rightly say that in the area of cost recovery these private waste stakeholders are effective and that is the reason the companies are being sustained up to the present time. In order to ensure sustainability of their operation, new entrants of the operators are at the moment being registered by the newly created Kano State Ministry of Environment. The operators can also get loan facilities from banks and other financial institutions in Kano metropolitan or even from the government because they are registered firms. It has been claimed that the private operators primarily cater to middle- and upper-class residents [16]. The old institutional setup may need to be changed for greater efficacy if they are to function well in the city's low-income neighborhoods. This alteration may take the form of private companies collecting and disposing of the solid waste from these locations; the state government, the local government, and the affected communities would then split the cost. The state and LGA should be in charge of developing and implementing the policy in order for this to be effective. The roles and duties of the stakeholders should also be clearly stated. The government, after giving a monopoly right to individual franchisees, also provides enforcement, adequate supervision and maintenance of disposal sites [17].

The above can be achieved by setting up of an effective community mobilization strategies whose responsibility is to provide awareness on the importance of sanitary habits; establish a system of cost sharing formula between the state/LGA/communities for the implementation; where to source for funds that can be used to accelerated implementation of the program me; contribute financial and material support to LGAs and communities; and facilitate construction and monitor implementation. The community's responsibilities as solid waste stakeholders will also include mobilization and motivation of community members on the need for good sanitary habits through solid waste hierarchy concept of on-site source recovery and recycling of waste, identify and cost the resources needed for solid waste management hierarchy and sanitation activities within area of operation; ensure that households contribute financially, provide material support to households and monitor implementation of the whole operation until success is finally achieved in both recovery, recycling, collection and final disposal of refuse. **Figure 1** shows proper recovery and recycling of MSW by the operators at the operators temporal waste storage facilities.

3.5 Auxiliary benefits for workers

Habitually, residential waste is composed with different kinds of recovery and recycling products such as plastic materials, bottles, and metallic materials. These kinds of products are being sorted by the employees of these private waste companies; normally laborers in all the areas of operation do sell them to either full term scavengers or scraps vendors [18]. And these are further resold to recycling companies for production of new items, both on-sorted and sorted materials from municipal solid waste are further separated to form a better recyclable that can be transported to bigger industries in Lagos and beyond by these stakeholders. However, the benefits from such sales of recyclables in the first instance above do not go to the private companies directly, but to their workers who greatly benefit from it. An in-depth investigation of one waste private contractor company reveals that, before disposal, its workers, who are also stakeholders, typically sort 35.1 kg of reuse, recovered, and recycled materials from the waste they collect in the operational area assigned to them each day on average. They make approximately =N = 49,440 per month and =N = 433,280 year from sales of these materials. The materials that are being sorted and their corresponding unit costs are displayed in **Table 7**, **Figure 2**.

Type	Amount (kg)	Price/kg (=N=)	Total =N=	Per month	Per annum
Bottles	6 3	60	378	11,340	136,080
Metals	11.2 5	5	6	1,680	20,160
Plastic	13.8	10	138	4,140	49,680
Polythene (HD)	3.8	20	76	2,280	27,360
Total	35.1	—	648	19,440	233,280

Source: Fieldwork, 2021.

Table 7.
Daily average sorted reusable materials and their value.

Figure 2
Recovery and recycling of MSW in ancient Kano city Nigeria. Source: Field survey 2021.

4. Outstanding issues

4.1 External conditions that affect the practice

There are some external conditions, which affect the practice. Over the years governments have known that for effective solid waste collection and disposal, the private sector will fare better. However for political reasons the state government has been directly involved in environmental sanitation even though without success. The Refuse Management and Sanitation Board (REMASAB), Environment Task Force on Sanitation, Kano State Environmental Planning and Protection Agency (KNUPDA) etc. have been created and scrapped. This attitude of the government has affected the private operator negatively [12]. They cannot be aggressive in expanding their area of coverage because of the government's direct involvement. Also more new entrants cannot be encouraged by this attitude of the government. Instead of seeing government action as complimentary, they see it as competing. This activity is badly impacted by some of the issues that have been happening in Kano City, especially if the operators are not natives. There may be religious or tribal crises [19]. They regard government action as competing with them rather than as a complement. This activity is badly impacted by some of the issues that have been happening in Kano City, especially if the operators are not natives. There may be religious or tribal crises [19]. The fact that the scope of their operations is limited to residential areas, even though those areas are inside their clearly defined operational grounds, is another persistent problem that all the private enterprises that were interviewed voiced complaints about. According to 55% of the respondents, the poor state of some of the disposal sites, particularly during the wet season, makes it difficult for vehicles to go through them and prevents them from operating fully. **Figure 3** contains a scenario showing the present condition of our dumpsite located with residential buildings couple with scavengers at work.

4.2 Information gaps during study

There are missing information gaps, which affect the analysis of the practice. Most of the private waste stakeholder's operators are secretive about their operation.

Figure 3
Showing the state of dumpsites at the study area. Source: Author field survey 2021 a negative recent advance in solid waste management.

In fact the researcher found it difficult to have access to the operators, throughout the research work granted audience once. And if the necessity for a meeting with the management arises, the researcher is informed that the manager is either unavailable or busy. This is because they view the researcher as someone who may suggest to the government measures that will drive out private operators from the market. When the researcher has access to information, the operators may provide general information. For instance, they were unable to provide the number and addresses of their clients. On the other hand, gathering information was difficult in communities of low-income earners in the city due to cultural norms, where the majority of the women to be interviewed were in purdah and the researcher needed an interpreter at the interview to interpret the questions, and in most cases, the husband had to be invited.

5. Lessons learned/conclusion

5.1 Lessons learnt

In spite of the ongoing successful journey of a private waste contractor and or private sector stakeholders in municipal waste in Kano metropolis is having some operational challenges which if adequately taken care of, the full benefits identified above would be fully realized. During this research study on challenges of SWM in terms of recovery, recycling, collection and final disposal of these waste by private SW stakeholders operators in Kano City, some issues have been identified as very important and these are:

- Urban communities are willing and able to contribute to the improvement of their areas in terms of sanitation on a cost-sharing basis through recovery and possible recycling of this waste. In other words, the community is willing to pay some percentage for waste recovery/recycling and or collection and disposal by private stakeholder's operators while the state or local governments will then pay the balance to the effective solid waste stakeholder operator.

- However this suitable program will require adequate and efficient institutional arrangement/ modification at all levels of implementation. Full participation of the private sector stakeholders in solid waste recovery, recycling, collection and disposal is highly welcomed since the government cannot cope up with the huge amount of waste generated as shown in **Figure 4** below.

- The state and municipal governments are working to educate the public about the importance of a clean environment. In this regard, the Honorable Chairman of the Nassarawa Local Government Area frequently appears on radio.

- Women and children are devoted to the effective operation of the waste management system and are proud stakeholders in this arrangement.

- As more private companies enter the solid waste collection and disposal industry, slum and peri-urban populations will be able to afford their services because competition is a key role in ensuring reduced rates for prospective customers.

Figure 4.
Showing the mountainous heap of solid waste at a dumpsite in Kano metropolis. Source: Field survey 2021.

5.2 Conclusion

It is obvious from the analysis done previously on private garbage collectors operating under a franchise agreement in Kano metropolitan that their business has a role in urban solid waste management. This leads to the conclusion that their involvement in waste management extends beyond successful recovery, recycling collection and disposal, and a decrease in governmental activities to include the growth of the macroeconomic and specific industries. Since franchising is not possible, it is advised that the high density sections of Kano metropolitan should also be managed by private enterprises under contract. The government should perform its duty of providing proper oversight and enforcement when in the franchised regions, especially to deal with illegitimate operators and fee defaulters. Additionally, the location of disposal sites needs to be improved to allow for easy vehicular entry and exit. Campaigns for public education and awareness should be supported both in the research area and across Nigeria. The amount of garbage produced would significantly decrease if every resident was aware of the effects their waste has on the environment and public health. Therefore, for the waste management system to advance, general public knowledge is necessary. SWM effects on the environment and public health must be addressed and evaluated in their context. A considerable sum of money has already been spent without a thorough appraisal or development strategy. Only a few towns were able to create an efficient mechanism for a door-to-door collection system due to ineffective government legislation. Solid waste storage is a challenge as well because collection trucks can't pick up garbage every day. The secondary bins are frequently seen to be overflowing with leachate, occupied by rag pickers, and engaged in conflict with other creatures like street dogs, cows, rodents, etc. Due to poor infrastructure, the solid waste transportation system is still in the stage of development. Education and enlightenment programs are key here and it would facilitate the development of technical know-how on the conversion of waste to resources thereby enlightening more people to embrace new technological strategy skills on areas in the management of waste. Noting the complex characteristics and conditions of the low-income settlements, community based programs involving community based organizations, nongovernmental organizations and other service providers are a necessary complement to utility managed sanitation services. The project ensures that partnerships are developed not only between utilities and communities but also between utilities and service providers. The private operators engaged in solid waste collection and disposal are the other service providers and therefore their operational modes and limitations can aid. The government still have a role to play in establishing a public partnership in the various phases of solid waste management to limit the over-dependence exerted on state or its agency for waste management and increase efficiency in the management process, implementing and enforcing SWM policies as well as developing efficient avenues for the acquisition of subsidized technologies that are needed in the waste management process.

Acknowledgements

Appreciation to REMASAB, ministry of environment in collaboration with the Kano State Urban Planning & Development Authority (KNUPDA), who help with the much needed information that this study required to be successfully completed.

Conflict of interest

The authors declare no conflict of interest.

Data and material availability

Information regarding this study is available by contacting the corresponding author.

Author details

Muhammad Saleh[1]* and Azizan Marzuki[2]

1 Faculty of Earth and Environmental Science, Department of Urban and Regional Planning, Bayero University, Kano, Nigeria

2 School of Housing, Building and Planning Universiti Sains Malaysia, Malaysia

*Address all correspondence to: salehmuhammad@studentusm.my

References

[1] Kumar A, Agrawal A. Current research in environmental sustainability recent trends in solid waste management status, challenges, and potential for the future Indian cities—A review. Current Research in Environmental Sustainability. 2020;**2**:100011

[2] UN Habitat. Solid Waste Management in World Cities, water sanitation in world cities. Conference. 2010

[3] Joshi R, Ahmed S. Status and challenges of municipal solid waste management in India: A review. Cogent Environmental Science. 2016;**28**(1):1-18

[4] Salami HA, Adegite JO, Bademosi TT, Lawal SO, Olutayo OO, Olowosokedile O. A review on the current status of municipal solid waste management in Nigeria: Problems and solutions. Journal of Engineering Research and Reports. 2019;**3**(4):1-16

[5] Royo S, Yetano A, García-Lacalle J. Accountability styles in state-owned enterprises: The good, the bad, the ugly: The pretty. Revista de Contabilidad-Spanish Accounting Review. 2019;**22**(2):156-170

[6] Bichi MH, Amotobi DA. Characterization of household solid wastes generated in Sabon-gari area of Kano in Northern Nigeria. American Journal of Research Communication. 2013;**1**(2009):165-171

[7] Ehizemhen IC. Communal solid waste problems in northern part of Nigeria: Proposed waste management, regulatory and knowledge resolution. 2021

[8] Mohammed Iliya EH. Strengthening the Capacity of Water Utilities to Deliver Water and Sanitation Services, Environmental Health and Hygiene to Low Income Communities. Case Study for Kano (town), Nigeria Context and Practices. WUP: WATER UTILITY PARTNERSHIP FOR CAPACITY BUILDING IN AFRICA Water, 5. 2000

[9] Agbesola Y. Sustainability of municipal solid waste management in Nigeria: A Case Study of Lagos. 2013

[10] Pires A, Martinho G, Rodrigues S, Gomes MI, Pires A, Martinho G, et al. Prevention and reuse: Waste hierarchy steps before waste collection. Sustainable Solid Waste Collection and Management. 2019;**2018**:13-23

[11] Adamu Mustapha ABN. Enhancing awareness and participation of municipal solid waste management in Kano Metropolis, Nigeria Aliyu. World Scientific News. 2014;**5**:26-34

[12] Umar UM, Naibbi AI. Analysis and suitability modeling of solid waste disposal sites in Kano metropolis, Nigeria. Geocarto International. 2020:1-19

[13] Lambu IB. Waste or Wealth' the Cultural Crux Behind Scavenging in Urban Kano State, Nigeria. 2016;**3**(3):19-25

[14] Nabegu AB. An assessment of Refuse Management and Sanitation Board Kano (REMASAB). Techno Science Africana Journal. 2008;**1**(June):1-15

[15] Wahab B, Ola AB. The mode of operation, challenges and benefits of informal waste collection in Ibadan Keywords: Informal waste collectors, mode of operation, challenges, benefits, Ibadan. Ibadan Journal of Sociology. 2017;**6**

[16] Mustapha ABN. Institutional Constraints to Municipal Solid Waste Management in Kano Metropolis, Nigeria. International Journal of Innovative Environmental Studies Research. 2015;**3**:13-21

[17] Naibbi AI, Umar UM. An appraisal of spatial distribution of solid waste disposal sites in Kano Metropolis, Nigeria. Journal of Geoscience and Environment Protection. 2017;**5**(11):24-36

[18] Ibrahim AM. The role of private sector participation in municipal solid waste in Kano Metropolis. 2014;**3**(5):37-42

[19] Winter M, Ujoh F. A review of institutional frameworks & financing arrangements for waste management in Nigerian Cities. Urban Studies and Public Administration. 2020;**3**(2):21

Chapter 2

Optimized Planning and Management of Domiciliary and Selective Solid Waste: Results of Application in Brazilian Cities (SisRot®Lix)

Marcos Negreiros, Augusto Wagner Palhano and Eduardo Reis

Abstract

We show a new technology to manage solid waste services through optimization methods (on sectoring, routing costs, and resources). This technology is called optimized planning and integrated logistics management (OPILM). It is being applied to Brazilian municipalities as it attends to their major natural features. The technology is formed by a framework of computational systems that uses optimization methods from sector arc routing and scheduling, fleet and staff scheduling, using also mobile smartphone apps. We present some of the results of real cases evaluated for residential refuse collection and selective waste collection in two Brazilian cities (Petrópolis/RJ and Bom Jesus dos Perdões/SP). The plan implementations achieved 17.9% from actual fixed and variable cost savings for sectors (vehicles and workers) and routes (time and distances) for residential refuse collection in Petrópolis/RJ. For the selective waste collection, we detail how we made our project to Bom Jesus dos Perdões/SP. We also present the returns considering costs involved in the management of the operational level and amortized by the investment required to use and apply the proposed technology for Petrópolis/SP.

Keywords: domiciliary waste collection, street sweeping, sector arc routing problem, spatial DSS, ERP

1. Introduction

Solid waste management deals with planning resources that are largely neglected by major Brazilian municipalities. The number of services related to the waste management in municipalities goes is more than 16. To mention some of these services, we have: domiciliary and selective waste collection (door-to-door), mechanical and human street sweeping, commercial and health waste collection, fair waste removal, cleaning and sweeping, container waste collection, painting gutters, and so on. All these services need plans and detailed establishment of the operation. Major work is done using unspecialized software tools, such as GIS-based software, but there is decision-making technology available, not well known by environmental engineers and managers, which can be suitable and adequate for better planning and optimizing resources [1].

For the domiciliary (door-to-door) waste collection, we define a sector as a defined region where staff (driver and collectors) do the domiciliary collection in a workday with the same weekly frequency. A circuit (shift, route, or collection trip) is the sub-region where the vehicle used by the staff travels until it reaches its waste storage capacity. A trip is a route between the garage, collection sector, and destination (typically a transfer station or landfill). Reinforcing the definition, a sector is composed of one or more collection circuits or routes that can be performed during a workday [2].

Most of the domiciliary and selective waste collection managers in Brazil plan their sectors and routes by using paper and drawing plans. The more effective and advanced ways of planning use AutoCad™, other tools incorporated by geoprocessors like Caliper™ [3], and apis from ESRI™, Here Maps, Open Street Map, Google Maps™ and others.

This work shows how we have solved these optimization and planning problems by using spatial decision support system (SDSS) technology composed of a software framework that allows users to design and manage mechanical collection and other service plans. To prepare the plans, we used SisRot® LIX, from Graphvs Ltda. [4], which calculates the collection sectors and street-by-street routes. The motion to use this system in counterpart to the others mentioned is that the SisRot®LIX is specialized in sector-routing problem, and it generality gives us back the opportunity to show the appropriateness of sector-routing for waste collection in different areas of waste collection management like domiciliary waste collection and domiciliary selective waste collection.

We present an application where we have re-sectored and rerouted the Petrópolis/RJ municipality, including the fleet, man hours, and overtime reductions. Finally, improvements are also presented for the new sectorization and optimization of selective waste collection sectors and routes in the city of Bom Jesus dos Perdões/SP, where the best placement of a recycle plant in the city was also studied.

We have organized this work into six sections. In Section 2, we illustrate the differences between capacitated arc-routing problems (CARPs) and SARPs. The third section presents the OPILM architecture with the SDSS SisRot® LIX features. Section 4 presents the results of the technology application. Section 5 studies the effect of savings in the waste collection system, taking as example the work done in downtown Petrópolis/RJ, here evaluated, and Section 6 summarizes the conclusions.

2. General sector arc-routing problem and its related problems

For better understanding the problems related to routing waste collection vehicles, we indicate below three of them:

Let us consider a strong-connected mixed graph $G(V\text{-}V_R, V_R, L\text{-}L_R, L_R)$, where V is the set of vertices, V_R are the required vertices, and L is the set of links ($L \subseteq E \cup A$, E—not oriented set of edges, A—oriented set of arcs, and $L_R \subseteq E_R \cup A_R$, E_R— set of required edges and A_R—set of required oriented arcs) that must be covered for the known garbage offer ($q_R > 0$). With this basic statement, the following problems consider the elementary needs to be solved by specialized methods [5].

There are three different methodologies of designing routes for waste collection: Chinese Postman Problem or CCP-based sectoring and routing, CARP-based solutions, and SARP-based solutions.

The CCP-based solutions consider that there are areas of collection that must be covered at certain times and on certain days. These areas can be covered in less than one workday through route design. The goal is to minimize the number of trucks required to cover the areas while joining (sectoring) different coverage areas in multiple trips to the landfill or transfer station. The routes are partially built by the Chinese Postman method, and a truck scheduling problem is performed to join the trips and compose sectors using a given fleet [6].

In the CARP-based solutions, the methods used consider a strongly connected positive-weighted graph as indicated previously. A fleet (homogeneous or heterogeneous) starts and ends at the garage ($v_0 \in V$) and passes through one or more intermediate facilities ($v_I \in V$), that is, a landfill or transfer station. This problem was first proposed as a symmetric graph $G(V, E\text{-}E_R, E_R)$ by Golden and Wong [7].

Using this approach, residential refuse collection applications and software for different cities worldwide were investigated by Ghiani et al. [8], and the most recent surveys on methods and their applications in garbage collection were conducted by Ghiani et al. [9], Han and Ponce-Cueto [10], and Mourão and Pinto [11]. With CARP solutions, the coverage of the area is not coordinated in a specific region, that is, shift contiguities are not guaranteed, or the same vehicle may collect in opposite places in a city, making managing the execution and measurement of daily processes complex [12]. Most recently, advanced methods have been developed by Boyaci et al. [13], considering fast lower bounds for CARP with intermediate facilities, and Janela et al. [14], considering balanced routes heuristics for MCARP.

The SARP was proposed by Mourão et al. [2]; it is considered an arc-routing problem (ARP), although it can be extended to node-routing applications for commercial and health waste collection. The SARP is, however, a more general problem than CARP as it includes all the problems of routing in arcs and nodes either with or without intermediate facilities [15]. In the SARP, trip contiguity, connectivity, and network compactness in a sector are considered objectives to be achieved by the final sector circuits covered by the routes, considering workload and fleet capacity constraints [16].

SARP resolutions are found by using clustering-based heuristic algorithms from generalized heterogeneous group assignments (sectors and circuits), and the Mixed Rural Postman Problem (MRPP) was applied to the circuits by Batista et al. [17] and Corberán et al. [18]. CARP heuristic method modifications have also been used by Corberán et al. [18]; they have been used by Ghiani et al. [19, 20] in zoning. Wølk and Laporte [21] used CARP with districting for waste collection in Denmark, and Boyaci et al. [13] applied fast upper and lower bounds for CARP with intermediate facilities and deadline methodology to large-scale world city instances. SARP was investigated by Mourão et al. [2], Rodrigues [15], and Cartinhal et al. [22].

In **Figure 1**, we can see an example of two circuits (trips) in the same collection sector: the chart on the left indicates the first trip (first circuit), starting at the garage, passing through the sector to collect, and ending at the disposal site. The chart on the right shows the second trip, starting from the end of the previous trip (transfer station), passing through the collection sector, and returning to the disposal site for unloading. The routes between garage and sector, sector and transfer station, and then transfer station and sector can be optimized by the "shortest path" or the "fastest path."

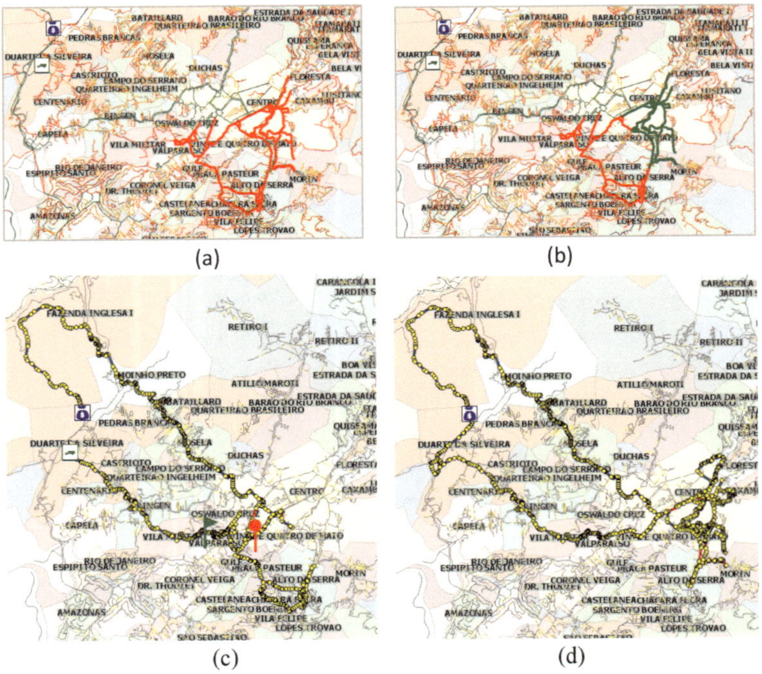

Figure 1.
Sector 220 and its circuits with routes for each circuit. (a) Sector 220 – Petropolis/RJ. (b) Circuits of the sector 220, (c) Sector 220 first trip, (d) Sector 220 second trip.

3. DSS-OPILM and SisRot® LIX features

The OPILM SisRot® LIX has been in development since 1989. Its scheme (**Figure 2**) presents the operational architecture of a set of computer systems that integrate through the web, allowing for optimized SWM (solid waste management)

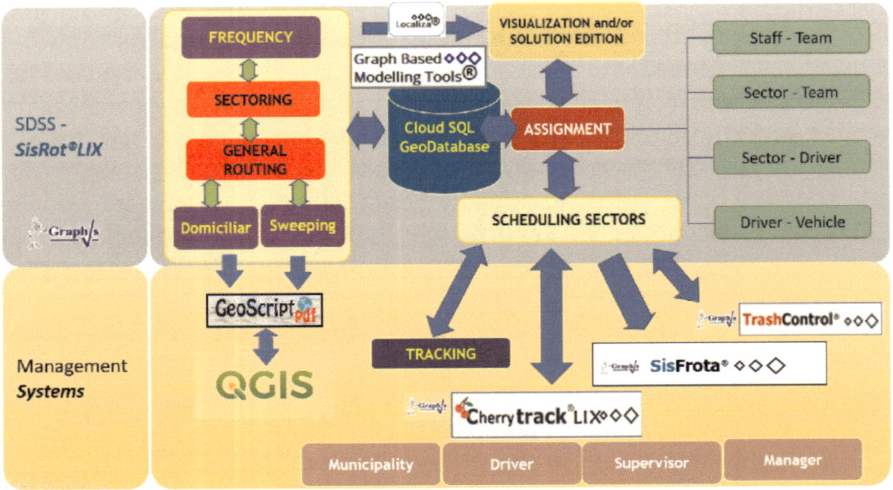

Figure 2.
OPILM architecture of the framework of computational systems for SWM [4].

and planning the different services to be performed. The system architecture is closely related to the conceptual DSS design model proposed by Klashner and Sabet [23] and the architecture proposed by Negreiros et al [24] and the web based framework developed to dengue control (webdengue) [25]. The core of the framework, corresponding to theory and analysis, includes the SARP system, which composes the algorithms for capacitated sectoring and routing, which are executed either manually or automatically. Heterogeneous sectoring is performed by heterogeneous capacitated centered clustering math heuristics, and the routing phase is performed with the mixed rural postman and movement prohibition reduction methods. These algorithms are executed according to the application (residential or selective domiciliary collection or human/mechanical street sweeping). The sectors and routes are part of the process final decision, and they are visualized through multigraph diagrams with movement/load/time/distance descriptions, multi-scaled vector maps (in different file formats), and street-by-street descriptive spreadsheets (in the XLS file format).

To better understand the OPILM in the portion of the SisRot®LIX, as shown above in **Figure 2**, a set of computational resources and parameters that govern the operational conditions in the field are necessary, as listed below:

1. Graph-based modeling system to network generation and editing, inclusion of constraints, editor connected with the associated roadmap shapefile (.shp)

2. Network-isolated vertices, strong connectivity, and number of components verification

3. Required street segment assignment and edition.

4. Definition of street features (Road, ave., street, etc), zip codes, numbers min and max of each street segment, and so on.

5. Editable load assignment spreadsheets per street segment (according to area population or proportional production per meter); the rate of waste generation per capita/region or by distance traveled (m); in the absence of this information by the operator SisRot®LIX being integrated into the Brazilian census map, which can determine the rate of waste generation by population density or by real estate

6. Inclusion and editing of special collection points and specific cargo (containers, fairs, accumulation points, customers)

7. Inclusion and editing "pulls:" there are roads taken by collectors, who collect the garbage up to the vehicle, either due to the physical limitation of the way the vehicle cannot cross it or because it has little or no waste; SisRot®LIX treats these roads for optimization, either disregarding them on the route (traffic impossibility) or using them (roads with little/no waste generation) if the calculated route is the most economical with traffic on them.

8. Editor to reformulate sectors and circuits

9. Location of the origin(s) and destination(s) of the vehicles: SisRot®LIX optimizes the operation with several garages and/or landfill/transfers (multi-origins and multi-destinations).

10. Slope of the road by using the spatial coordinate height (possibility to use generalized windy rural postman problem)

11. Vehicle load capacity (mixed fleet: vehicles can have the same or different loads)

12. Duration of the team's workday (household collection, manual/mechanized sweeping, and collection of swept garbage)

13. Vehicle speeds empty, full, collecting or deadheading

14. Collection discharge time at the landfill or transfer station

15. Edition of forbidden turns

16. Paths with/without forbidden turns

17. Mixed node and arc-routing methods (Generalized Rural Postman Problem in mixed networks) [16]

18. Multiple collection in arcs (special streets) in different trips.

Some characteristics of the solutions:

1. It is possible to optimize the selective and punctual selective collection routes (eco points, containers, bags resulting from sweeping, street fairs, monuments, hospital, etc.).

2. Vehicle routes are generated with a minimum of "U" returns and maneuvers to the left or right: less execution time and less effort from the driver, as well as keeping the service as continuous as possible.

3. Balanced trips: the routes can be optimized so that the loads collected between trips in the same sector are close, favoring the reduction of total workforce time.

4. Real-time tracking/visualization of vehicles, "lutocares," and any equipment and comparison to what was planned.

5. The routes between garage and sector, sector and disposal, and disposal and sector can be optimized both by the "shortest path" and the "fastest path." The section of the route that comprises the collection service itself, within the sector, is always optimized by the "shortest path."

6. The sweeping bags can be programmed by the sweeper every "*x*" meter and always close to corners. The route of the vehicle that will collect the sweeping bags, as well as the route of the vehicle used to take and pick up the sweepers for the sectors, is also optimized using a node-routing system, SisRot®FULL [4].

7. "Lutocar" tracking: any tracker can be incorporated into the system, allowing to monitor the execution of the sweeping service in real time. Any other equipment can also be tracked and incorporated into the system.

8. Assignment of teams of collectors to sectors, vehicles to sectors, and drivers to sector.

9. Zone-to-sector frequency definition and assignment.

10. Visualization and map/descriptive reports of zones, sectors, circuits, and frequency in different formats (KML, PDF, XLS).

11. Visualization and map/descriptive reports of routes in different formats (multigraph, animated, KML, PDF, XLS, and many others).

There is other arc routing software available in the market like TRANSCAD® [3]) and RouteSmart® [26] (from RouteSmart Technologies™). As can be seen in Environmental Expert site [27], there are many other platforms for waste management, analysis, monitoring, and planning using sensors (Smart sensors, watchdog, RFID tags, smart button) using specific modules for asset management, waste monitoring, route planning and fleet management from Sensoneo™ [28], and route planning from EasyRoute™ [29]. This late software are not arc routing based, they were built to accomplish point-to-point bins in domiciliary waste removal.

4. Designing and monitoring of sectors and routes

In this section, we consider the two types of results we obtained by using the OPILM SisRot LIX©. The residential refuse collection and domiciliary selective waste collection results are detailed in the following subsections.

4.1 Household refuse collection

With a population of 307 thousand inhabitants [30, 31], Petrópolis-RJ is a well-known touristic city; it has 48 collection sectors with an average of two collection circuits per sector. Five sectors were performed by the concessionaire downtown. The June 2020 data (without optimization) of these sectors were informed by the managers of the local concessionaire as follows:

1. Sector identification: Sectors 220, 221, 222, 223, and 224 (**Figure 3** left)

2. Collection through 9-t "stump" compactor truck with a maximum collection load of 9.3 t for each trip can be used per sector

3. Collection frequency: night and daily (Mon-Sat)

4. Each sector required two trips (circuits) per vehicle to be fully covered, with the first trip being garage→sector→transfer station and the second trip being transfer station→sector→transfer station.

5. Average $\pm$ Std of garbage total production per sector (five sectors): 12,34 $\pm$ 3,63 t/sector (Mon: 15,56 $\pm$ 4,21 t/sector) and (Tue-Sat: 11,64 $\pm$ 3,08 t/sector)

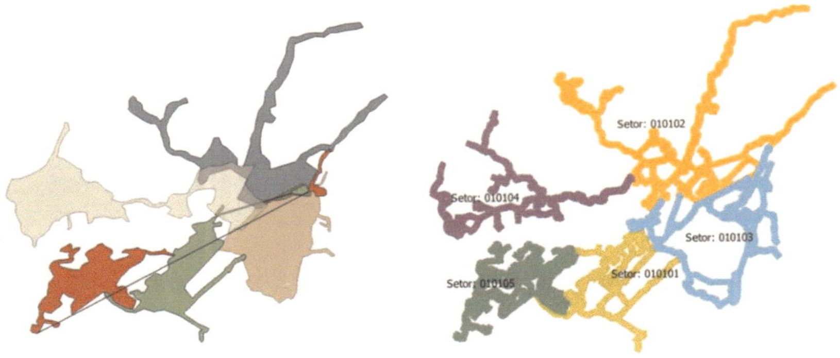

Figure 3.
Maps indicating the division of operation in downtown Petrópolis-RJ, considering at right previous sectoring (overlapping) and new coverage after project's intervention.

6. Total average daily distance traveled by fleet: 455 km/d (Mon: 465 km/d) (Tue-Sat: 450 km/d)

7. The average balance between daily trips: 90.5% (Mon: 9279%) and (Tue-Sat: 89.52%)

8. Team: five drivers and 15 collectors operate (one driver + three collectors on average in each sector)

9. Average ± Std collectors' efficiency (during collection): 823,02 ± 252,74 kg/man-h (Mon: 917,03 ± 312,85 kg/man-h) and (Tue-Sat: 802,35 kg ± 232,99/man-h); three collectors on average.

10. Collectors' average ± Std efficiency – with three collectors: 849,46 ± 211,01 kg/man-h; with four collectors: 581,59 ± 256,28 kg/man-h.

Figures 3–7 present the planning process to rearrange actual sectors and routes for the concessionaire. The development of the plans considers the unique precedence of

Figure 4.
Maps indicating the division of operation on Mondays in downtown Petrópolis-RJ, where on the left, we see the original five sectors (non-overlapping) and on the right, the final optimized sectors.

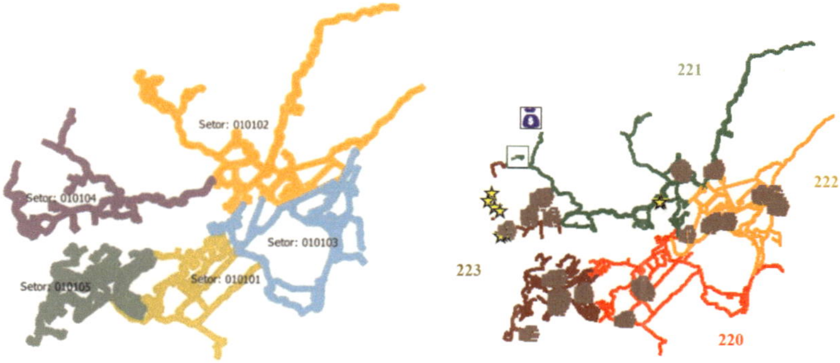

Figure 5.
Maps indicating the division of operation from Tuesdays to Saturdays in downtown Petrópolis-RJ, where on the left, we see the original five sectors (non-overlapping) and on the right, the final four optimized sectors for these days.

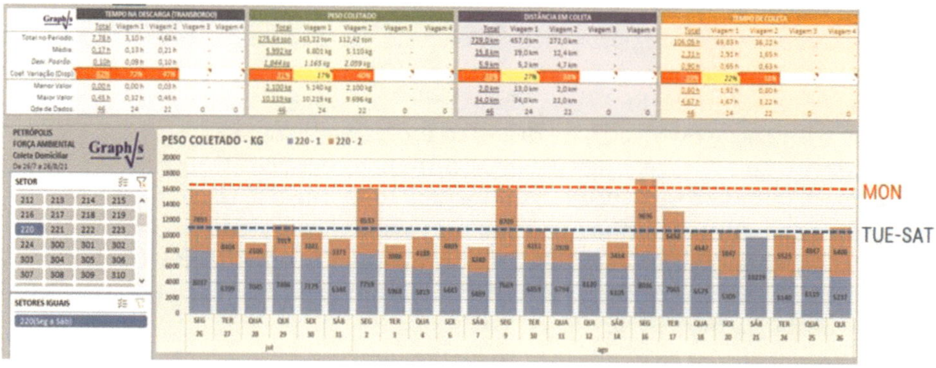

Figure 6.
Sector 220, high variability production between Mondays and Tuesday–Saturdays (more stable).

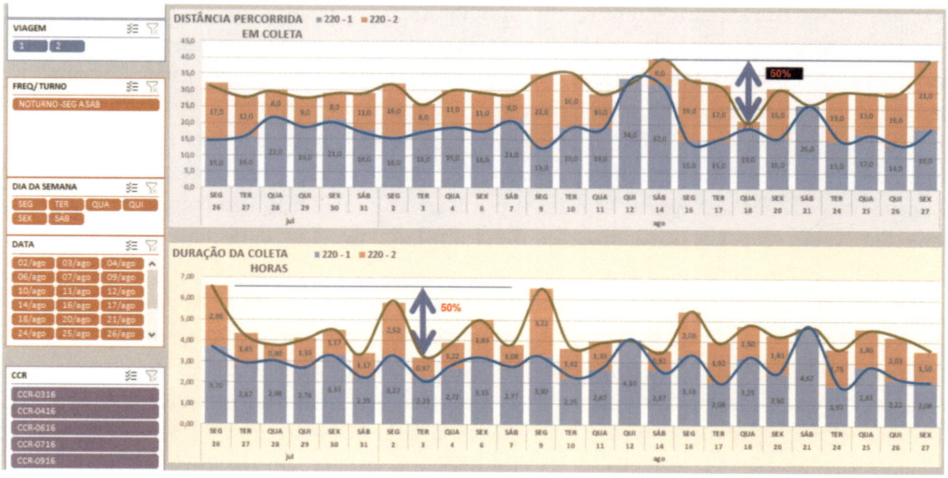

Figure 7.
Sector 220, above the variability between total distances traveled per trip (above) along the days of the week and variation in time duration (below) between Mondays and Tuesday–Saturdays.

the garbage (no sector overlapping) and minor changes between days of collection to warranty the monotonicity and completeness of the service.

Figure 3 presents the state of the planned sectors we found in downtown Petrópolis-RJ; they were covered daily at night. The sectors were planned considering overlapping coverage (**Figure 3** - left), meaning that it was no longer possible to precisely identify the places where the garbage came from. It was necessary to separate the groups to be unique and make the statistics of production (**Figure 4**-right).

Figure 4 presents the previous (left) and new sectors' (right) configurations for the new spatial division in reason of the optimization of the sectors and routing to be performed on Mondays. **Figure 6** presents the previous (left) and new sectors'(right) configurations for the new spatial division to be performed from Tuesdays to Saturdays with four vehicles only. It can be also seen in **Figures 4** and **5** that the collection is also performed in star points (containers, gas stations, bars, supermarkets, etc) and brown points (called "manual collection"—places where the vehicle cannot come in and/or traverse. It normally stops, and a collector take it manually hundreds of meters away from the parking point).

The practiced routes are performed following a great route between all street segments from garage to transfer station. It means that the driver follows the sequence defined by the great route, and every time the driver wants to discharge (in general because of the vehicle overload), he drives to the transfer station.

In **Figures 6** and 7, we can see the great difference between the garbage production, distance, and time on collection for sector 220 by the city on Mondays (due to higher flow of commerce on this day) and the rest of the week. It achieves about 50% of the production between these days in downtown Petropolis-RJ, and it indicates that the service plans must be distinct for each period (Mon and Tue-Sat). For Tue-Sat, the dispersion of the garbage is very low.

Table 1 shows the scenario of the production practiced in each sector before the optimization. It is convenient to observe more precisely the variations between each sector on Monday.

Our field evaluations had been done in just one scenario, in case of the service movements realized for Tuesday–Saturday, because it may be extended for Monday, after adjusting the street coverage and special local constraints needed to be satisfied (one-way streets, passing more than once in particular streets, passing particular places in a specific time, manual collection, and others). The project identified all the turns and prohibitions during a period of 35 days.

After identifying unfeasible movements, many of them were lost because they were not properly identified. We prepared the drivers in the use and navigation of the CherryTrack®LIX for a week, and then we started evaluating the routes in the next week. We tried four times to run the plans in plain use. The first time, we evaluated the completion of the routes and use of CharryTrack® Lix by the drivers. Problems with the constraints (turns and prohibitions) and with the response of the app were detected that disturbed the execution of the routes, leaving many streets without coverage. On the second try, we had problems with changes in the field not being reported, vehicle overload before expected, and still matters with the app response. On the third try, we had problems with vehicle overload before expected and no more problems with the app response, and finally, on the fourth try, we only had problems with tire flat and no more, although a new change was necessary in the routes asked on the last day. The evaluations were done from October/2021 to January/2022.

DOI: http://dx.doi.org/10.5772/intechopen.111432

Sector	PRODUCTION							Km			HOURS			Bal	Discharge - t		
	Day	Trips	# Days	Total	Kg/Trip	Average (Kg)	Var %	Total	Ave	Var %	Total	Ave	Var %	Ave %	Indicated	Average	Var %
220	MON	8	4	66,334	9636.29	8291.63	8.11%	405	101.25	5.43%	32.58	8.15	5.16%	95.01%	7.92	8	0.95%
221	MON	8	4	68,546	10003.43	8568.14	8.38%	400	100.00	6.00%	30.17	7.54	3.78%	94.86%	9.95	10	0.45%
222	MON	10	4	76,132	11983.61	7613.19	28.70%	483	120.75	19.40%	34.03	8.51	12.97%	84.87%	8.91	9	1.03%
223	MON	11	4	79,595	10140.39	7235.89	20.07%	398	99.50	31.87%	36.40	9.10	20.87%	89.01%	6.96	7	0.61%
224	MON	2	2	13,621	8032.16	6810.11	8.97%	97	48.50	1.46%	10.53	5.27	34.91%	100.00%	10.79	11	1.93%
	MON	39	18	304,228	10001.56	<<– Ave Day/ Sector	20.65%	1783			143.72			92.75%			
	TUE-SAT	181	100	1,163,225	11632.25	<<– Ave Day/ Sector	7.79%	9195			651.03			89.52%		9.28	

Table 1.
Scenario of the production per sector for Monday and Tuesday–Saturday in downtown Petropolis' sectors.

4.1.1 Comparison between original situation and optimized scenario (Mondays)

The final solution obtained for Mondays is the one shown in **Table 2**.

As a result, we obtain the following if we consider the same number of days covered per sector (if we consider days: 16/08, 9/08, and 27/07):

- Little increase in total distance: before – 483 km, planned – 484 km (increase: 0.21%)
- Reduction in total route time duration: before – 124 h, planned –132 h (savings: 6.2%)
- Increase in balance between trips: before – 88.51%, planned – 94.23%

The difference in distance is because of the new constraints that are not practiced by the concessionaire are here considered in the final planned routes. The reduction is in fact because of the better-balanced routes take less time spent for collection in 6.15% as the plan established.

4.1.2 Comparison between original situation and optimized scenario (Tue-sat)

In the computation of the sectors and routes for daily night collection in downtown Petropolis/RJ, we needed to reconsider the previous collection sectors and the production statistics for Tuesdays-Saturdays. There were plenty of

Sector	Production-Kg	Distance (Km)	Time
220	48,176,64	286,155	26:18:30
221	50,391,69	307,491	26:32:00
222	49,657,23	351,486	25:15:12
223	45,394,77	185,598	19:38:09
224	47,907,27	321,246	26:01:36
Total:	241,527,6	1451,976	123:45:27

			DAILY NIGHT		
DAY	DATE	Km	HOURS	PROD	BAL
SEG	16/08/2021	505,00	46:40:00	86,726,64	80,99%
SEG	09/08/2021	491,00	43:36:00	78,242,52	90,77%
SEG	27/07/2021	453,00	41:36:00	75,077,97	93,75%
	Trips:	31			
	Total:	1449,00	131:52:00	240,047,14	

-0,62%	-0,21%	6,15%
Production	Km	Time

Table 2.
Solution obtained after all corrections in the field, to be applied to the sectors on Mondays at downtown Petropolis/RJ.

information for these cases, and we could build the routes with much greater success than before.

With the considerable decrease in the daily production between Mondays and the rest of the week as shown in **Figure 7**, the sectors were recalculated with the adequate parameters of production. We found a new configuration with four sectors. Running up the capacitated clustering according to the vehicle's technical capacity in the sectors to build the circuits, meaning each trip, as can be seen in **Figure 1c** and **d**, the solution achieved impressive savings in the first plan produced. Considering that many local constraints were not satisfied, we included them in the graph and obtained a new plan for the first execution in the field. As mentioned previously, the process continued until both constraint satisfaction and app return were in order; we tried four times, until the last, although still with problems with vehicle's breakdown, it meant that the completion was not compromised at all. **Table 3** shows the last result obtained, **Table 4** presents the plan produced by SisRot® LIX, and in **Table 5**, we present the estimated savings considering the numbers from production spreadsheets (92 trips, 27/07/21–29/01/22) with vehicles' breakdown during service and without this occurrence.

The savings achieved 17,86% with distance. The final plan also reduced 1 vehicle and 1 driver in 5 for both resources. We considered the effect of unemployment, and we maintained the same number of collectors, but it caused prejudice to the process, because the collection with three men was faster (9.54%) than a collection with four men. It means that on average a team with three men collects 2548.38 kg/h, while a team with four men collects 2326.36 kg/h in Petropolis/RJ. In **Table 6**, we can see the average efficiency (kg/man-h) of the collectors between sectors and days of the week to better understand this important variable in decision-making.

The experience with CherryTrack® LIX was fundamental in the process of establishing the sectors and routes in the field. After training drivers for a week, we had big problems with those who did not repass the training of studying the app at home. The only one who did it was successful in every route he did and always finalized his job before all. He was used as an example for the others, and three of them succeeded in the end. One more resistant did not want to use the technology and always avoided it. In the end, the supervisor and all drivers were fully integrated with the app and did not want to look at maps anymore. The possibility to control all the drivers anywhere, even driving a truck, was considered the most advantageous part of the app for the supervisor, and for the drivers, it was the route indication while in operation and the state of completeness of the service any time.

DAY	SECTORS					AVERAGE	
	220	221	222	223	224		
MON	917.07	1117.49	1078.38	869.83	1100.95	917.43	
TUE	922.14	913.80	1009.53	895.69	464.08	900.24	882.35
WED	846.94	988.05	1041.91	796.48	442.76	858.43	
THR	906.83	960.65	1043.30	848.87	442.66	893.46	
FRI	928.20	1030.42	927.28	780.95	558.33	839.28	
SAT	887.13	925.72	994.71	847.21	435.22	900.24	

Table 3.
Result of the application in the field, for the 4th revision plan to Tuesday–Saturday.

EXECUTED OPERATION- PETROPOLIS-FORÇA AMIBIENTAL - 01/29/22											
	Vehicle	TARE	DRIVER	No Collectors	Tp	Weight	L	Kg/Km-h	Kg/hom-h	Total Time	Distance
220	8656	11,270	IGOR	5	1	7540	67,20	45,50	245,69	4:55:00	55
	4G25	11,640	IGOR	5	2	8685		144,35	548,55	4:13:00	51
221	4G20	11,590	ANTONIO	4	1	4990	58,80	19,99	129,95	2:30:00	62
	4G20	11,590	ANTONIO	4	2	7170	29,40	40,63	253,94	4:17:00	52
222	4G16	11,510	ELCIO	5	1	7790	35,28	39,34	173,11	3:45:00	56
	8629	11,340	ELCIO	5	2						
223	8656	11,270	WALTER	4	1	5368	39,88	632,60	1581,51	7:30:00	113
	8656	11,270	WALTER	4	2	2610	100,80	15,92	91,53	4:32:00	54
						43,953	331,3621	134,05	432,04	42:00,0	443

Table 4.
The result of the 4th revision plan to Tuesday–Saturday (the time to complete the task is overestimated).

SECTORS							
Sector	**Trips**	**Crew**	**Weight (kg)**	**Bal%**	**Perimeter (m)**	**Km**	**Time**
220	2	4	16,863,59	97,90%	29,559,00	102.896	09:30
221	2	3	16,107,36	97,80%	35,726,00	102.417	10:17
222	2	4	16,552,51	99,47%	34,745,00	122.193	11:01
223	2	4	15,135,84	96,46%	27,064,00	104.909	09:00
			64,659,30	97,91%	127,094	432,42	39:49:00
				Coverage		**TUE-SAT (km)**	
Executed with no Breakdown				96,17%		506,32	
Proposed in 02/02/23				100%		432,41	
Savings:						14,59%	
				Coverage		**TUE-SAT (km)**	
Executed with no Breakdown				100%		526,46	
Proposed in 02/02/23				100%		432,41	
Savings:						17,86%	
CREW PROPOSED PLAN							
SECTOR						**TUE-SAT**	
220						1 M + 4C	
221						1 M + 3C	
222						1 M + 4C	
223						1 M + 4C	
224							

Table 5.
Result in savings considering service with and without vehicle breakdown.

Types of vehicle speeds	Speed (km/h)
Empty	45
Full	35
In collection	5
Deadheading	17

Table 6.
Speeds of operation considered in the project for the city.

We applied the same SARP methodology explained above in domiciliary waste collection to other Brazilian cities like Franca/SP, Campo Grande/MS, Mazagão/AP, and Búzios/RJ with success, obtaining 12–28% savings in the total distance traveled and from 5 to 9% savings in the staff's working hours. The relevance of the methodology opened new opportunities to make the waste collection more predictable and less susceptible to the variations of the load per day.

4.2 Selective waste collection

We developed a project in June 2018 for domiciliary selective waste collection for the city of Bom Jesus dos Perdões – SP. The city's population is estimated to be 24 mil inhabitants [32], and it is a small and winter/summer town. The project was specific to generating sectors and routes, and their schedule was for periods of one week, for a contractor [33] of the local selective collection association. The contractor implemented a new recycling plant for the association, with the support of the best Brazilian knowledge in recycling tools and operational management. Here we detail how we developed this project.

The geographical data set came from OSM (Open Street Map), Google Maps™, and Bing™, and it were adjusted with QGIS 2.18 to be completely legible and useable for sector-arc routing. The data from population density and distribution along the municipality were obtained from the "Setores Censitários, IBGE" [34].

The work consisted of creating sectors and collection circuits from the collection area, **Figure 8**, and studying a better place to install the recycling plant (Distrito Industrial or in the Transfer Station), respecting the data provided by the contractor. The sectors and routes were divided into three geographic levels of grouping areas (regional, zones, sectors, and circuits) and thus named (Regional: 01; Zones: 01..05; Sectors: 01..04; Circuits: 01.. 04). This method allows the total route of the vehicle to carry out the selective collection to reach the shortest/fastest possible length while at the same time allowing the most adequate execution and inspection on the field.

A sector in the domiciliary selective collection has the same definition as in the household refuse collection, and the trips represent the circuits. The basic difference is that the trips are composed of lighter loads, and the vehicles used are not compactors. The garbage is measured in volume, and it is disposed of in a way to be separated manually faster. The garage and the recycling plant were running in different places; the garage were in the Distrito Industrial but for the project evaluation it could also be considered in the same location of the recycling plant, as can be seen in **Figure 8**.

The plans considered operational speeds, as exposed in **Table 6**, where these values were obtained from other experiences in a city with same features. Basically, it is the defined speed while vehicles are empty, in the collection, deadheading (traversing between two collection points without service), and in full charge or going to the recycling plant.

The effort of the drivers was considered, as proposed by McBridge [35] and Bodin et al. [36], to reduce "U" turns in arc-routing route calculations (**Table** 7); we used the

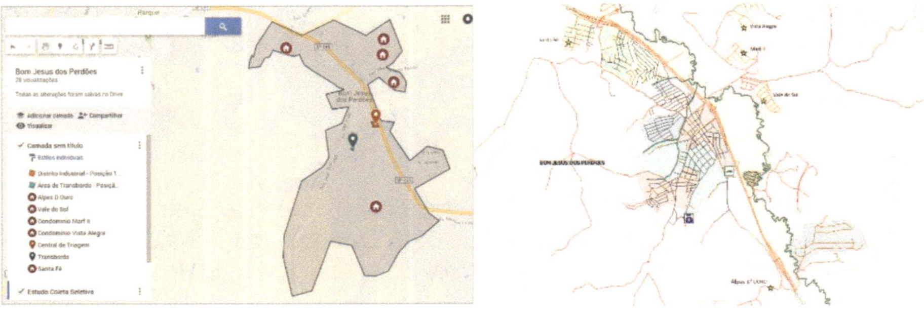

Figure 8.
Maps of the selective collection situation in the city of Bom Jesus dos Perdões on Google maps™ and SisRot LIX®.

Maneuver	Effort measure
Straight on	1
Turn left	10
Turn right	25
U turn	100
Service in adjacent links	Add 0 to the effort
Alternate service and no service in adjacent links	Add 5 to the effort
No service in adjacent links	Add 10 to the effort

Table 7.
Measures of steering effort in route movements.

approximative line-graph methods proposed by Negreiros & Palhano [37] to reduce the effort on maneuvers.

The workload was set to be 8:00 with an interruption of 1 h for lunch, as per the regular Brazilian labor laws. On tough days, like Mondays, when the garbage is of higher load than the rest of the week, there is a necessity of extra labor time; the association regulates this once all the people are in cooperation. The discharge time was defined to be 20 min because, in this case, there was a necessity for other people to come and help discharge manually the incoming vehicles.

The parameters related to the production of recycling were: 1 kg/inhab, where 30% were recycled with 55% of it recovered, totalizing 0.165 kg/day-inhab of recycled available (1 kg × 0.3 × 0.55). The collection would be performed by one vehicle with 28 m^3 of maximal occupation and 95% (26.6 m^3) – a surcharge of 20% is admissible. The admitted ratio between production and volume was 36 kg/m^3. The load collected by the vehicle per trip was 28m^3 or 957.6 kg (28 m^3 × 36 kg × 0.95) or at most 1200 kg or little more (±5%) in cases of possible seasonality.

- The city has four condominiums: Alpes D'Ouro, Vale do Sol, Vista Alegre, Santa Fé and Marf II, where the waste is more frequent on weekends and holidays. For them, it is admissible that the vehicle can carry kg.

The sectors were created to be covered one day a week from Monday to Friday. As can be seen in **Figure 8**, the polygon returns:

- The collection area: delimited by the polygon in gray.

- Recycling plant: where the vehicle may discharge.

The street network was identified, as in **Table 8**:

Street segments	Total	Street segments requiring service	Total collection perimeter (m)	Estimated total weight/volume total per week (−20%)
Doble way	403	124	63.353	37,936 kg
One way	4327	931		1354,86 m^3

Table 8.
Numerical data of the network and waste production estimates.

4.2.1 Census sectors', occupation, and production of the streets

Census sectors are defined by georeferenced polygons, where a census taker performs his survey task in the demographic census every ten years. The publicly available information informs a digital cartography of the number of households and residents in the inner region of the polygon that represents it. Using geoprocessing, the production of waste in each census sector is found by multiplying the number of people by the estimated production of selective waste per person per day of collection (0.165 kg/day of collection). By doing so, the perimeter of the street network considers only the places where the truck must pass (required street network) and receive the garbage distribution for each street segment, thus transforming the garbage in kg/m in the required graph. In the example in **Figure 9**, the required graph is weighted with 0.165 kg/person × 603 people = 102.79 kg, and given that this required graph perimeter is 1821 m, then we find 0.056 kg/m to distribute among the 1821 m of required connections of the set.

4.2.2 Collection zones

Table 9 presents the numerical data referring to the collection zones built with SisRot® Lix. **Figure 10** shows a geographical representation of the division of selective collection zones for better visualization. It is understood that this final configuration is the most appropriate, as it allows for better contiguity and adequate compaction, given the presence of the D Pedro I road, or Marginal Bom Jesus, separating the municipality into two large regions. It can be seen in **Figure 9** that to the north, the Vista Alegre condominium exceeds the city limits, causing an unfavorable impact on the calculation of routes in this zone.

Figure 9.
Census sector, reproduced from the digital base of the IBGE census sector mesh (IBGE, 2016) of the city of BJ dos Perdões/SP by SisRot® LIX. The required road network (in darker yellow) that is within the sector receives the production of the corrected population of the sector.

ZONES	No. street segments	Perimeter (m)	Production (kg)	Volume (m^3)	Sectors	Circuits	Vehicle (kg)
101	271	14,880	7940.9	220.58	2	8	1000
102	233	13,722	5663.18	157.31	2	6	1000
103	119	9072	8711.05	241.97	2	8	1000
104	240	20,612	7897.31	219.37	2	8	1000
105	191	12,687	7699.31	213.87	2	8	1000
Total (5)	1054	70,973	37911.76	1053.10	10	38	

Table 9.
Numerical construction details of the zones, with information on the required street segments.

Figure 10.
Map containing the division of the clusters of selective collection zones for the city of BJ dos Perdões.

4.2.3 Collection sectors

Table 10 presents numerical data referring to the collection sectors built with SisRot® Lix. **Figure 11** shows a geographical representation of the division of selective collection sectors for better visualization. This final configuration is the most suitable for the context of the local knowledge that has been made available. The dispersion of contiguous areas is noted in **Figure 11**, but all represent the same collection context, mainly in the condominiums to the northeast of the city (Vista Alegre, Marf II, and Vale do Sul).

4.2.4 Collection circuits

We finally arrive at the division of collection circuits. At this stage, it is necessary to guarantee the maximum contiguity between the circuits to compose the collection sectors and hence allow the creation of routes with the best operational results. **Figure 12** shows a geographical representation of the division of selective collection circuits for better visualization. This final configuration is best suited to the context of

Sectors	No street segments	Perimeter (m)	Production (kg)	Volume (m^3)	Circuits
10,101	136	6244	4006,6	111.2944	4
10,102	135	8636	3934,3	109.2861	4
10,201	78	6056	2623,18	72.8661	4
10,202	155	7666	3040	84.4444	3
10,301	43	4200	4421,88	122.8300	3
10,302	76	4872	4289,18	119.1439	4
10,401	136	13,219	3929,46	109.1517	4
10,402	104	7393	3967,85	110.2181	4
10,501	99	6948	4048,88	112.4689	4
10,502	92	5739	3650,43	101.4008	4
Total (9)	918	64,729	33,905,16	1053.1040	38

Table 10.
Numerical data of the selective collection sectors projected for BJ dos Perdões.

Figure 11.
Map with the division of sectors in the city of BJ dos Perdões.

the local knowledge that has been made available. Note in **Figure 12** the dispersion of contiguous areas further to the northeast (in condominiums).

Table 11 presents the numerical data referring to the collection circuits built with SisRot® Lix. Note that the nominal values allowed for each circuit were exceeded considering the maximum volume of the vehicle trunk of 26 m^3. At first, it should be considered that the updated IBGE data indicate 20% more in the population that was passed to this study; in addition, the seasonality of the places in overload brought the need to maintain the structure in this way. It is recommended that these areas be observed during the implementation phase of the routes, as there are no reliable data for this sizing.

The solutions presented in **Table 11** consider the following nomenclature:

Circuit: is the identifying label of the circuit (Reginal, Zone, Sector, Circuit);

Figure 12.
Visualization of the distribution of the collection circuits designed for BJ dos Perdões.

SSeg: Number of street segments in the sector;
Perimeter: is the total length of the required street segments of a circuit;
Prod: is the estimated production (in kg) of the circuit;
CDist: is the distance that the vehicle would travel per trip;
CTime: is the time spent on the trip;
Bal: is the rate between trips in the sector – Bal = $\left(\frac{\sum_{i=1}^{nv} Prod_i}{nv * \max\{i=1,..,nv{:}Prod_i\}}\right)$;
Load: is the estimated total load collected in the sector;
Time: is the estimated total travel time in the sector (hh:mm:ss);
Km: is the total distance traveled in the sector (in km);
Vehicle: is the maximum capacity—in kg (−20%)—of the reference vehicle.

4.2.5 Routes of collection circuits

All the circuits clustered were calculated for both final destinations in evaluation (Distrito Industrial and Transfer Station); these two places already have the basic infrastructure to install the recycling plant, and it was necessary to define the best solution obtained with routes and time between both. The result presented in **Table 11** refers to the installation of the plant in Distrito Industrial; the variable cost gained with this installation was 16.5% better than if it was installed in the Transfer Station.

An example of the multigraph of the routes directly extracted from SisRot®Lix can be seen in **Figure 13**. The use of the generalized sector-arc-routing solver that joins node and arc routing in the same context of decision-making is shown.

4.2.6 Sectors' schedule

Only one vehicle and one crew (driver and three collectors) will be needed to carry out the work of the entire city. The scale of the vehicle and crew, as shown in **Table 11**, indicates that the sector that covers the Alpes D'Ouro condominium (Sector: 010401) should be visited on Wednesday afternoon, since it is not possible to serve all the condominiums on the same day as required by RECICLEIROS. The Santa Fé

Circuits	SSeg	Perimeter	Prod (kg)	Vehicle	CDist.(km)	CTime	Sector	BAL %	Km	Time	Schedule	
1,010,101	32	1836	959,13	1200	6.019	01:02:17	10,101	93,54%	23,56	3:58:34	Morning	MON
1,010,102	49	1871	947,75	1200	5.766	01:01:27						
1,010,103	33	1353	1070,87	1200	6.546	01:01:08						
1,010,104	22	1184	1028,86	1200	5.224	00:53:42						
1,010,201	30	2113	1231,09	1200	5.242	00:59:43	10,102	79,89%	20,53	3:59:13	Afternoon	
1,010,202	45	2409	910,37	1200	5.470	01:03:07						
1,010,203	31	1979	1030,62	1200	5.378	00:59:48						
1,010,204	29	2135	762,22	1200	4.438	00:56:35						
1,020,101	28	2849	936,21	1200	14.431	01:49:13	10,201	91,34%	33.861	4:27:28	Morning	TUE
1,020,102	25	1424	729,71	1200	9.386	01:16:19						
1,020,103	25	1783	957,27	1200	10.044	01:21:56						
1,020,201	13	573	1119,07	1200	12.030	01:23:28	10,202	90,52%	44.186	5:22:38	Afternoon	
1,020,202	34	1116	1119,5	1200	12.371	01:29:04						
1,020,203	108	5977	801,42	1200	19.785	02:30:06						
1,030,101	12	1157	1066,93	1200	7.149	01:03:32	10,301	82,78%	29,79	4:17:28	Morning	WED
1,030,102	11	1369	837,31	1200	6.924	01:04:28						
1,030,103	9	811	1182,16	1200	7.344	01:02:19						
1,030,104	11	863	1335,48	1200	8.377	01:07:09						
1,030,201	15	1100	1233,79	1200	5.546	00:54:58	10,302	86,91%	27,14	4:05:53	Afternoon	
1,030,202	26	1661	1175,09	1200	7.120	01:06:02						
1,030,203	15	789	1006,29	1200	6.651	00:58:29						
1,030,204	20	1322	874,01	1200	7.821	01:06:24						

DOI: http://dx.doi.org/10.5772/intechopen.111432

Circuits	SSeg	Perimeter	Prod (kg)	Vehicle	CDist.(km)	CTime	Sector	BAL %	Km	Time	Schedule	
1,040,101	23	2141	952,97	1200	4.345	00:55:35	10,401	92,32%	28,33	4:57:47	Morning	THU
1,040,102	21	1758	940,27	1200	3.562	00:49:25						
1,040,103	77	7770	972,14	1200	16.992	02:26:29						
1,040,104	15	1550	1064,09	1200	3.429	00:46:18						
1,040,201	18	1646	1067,05	1200	3.751	00:49:05	10,402	92,96%	18,97	3:39:56	Afternoon	
1,040,202	28	2182	914,29	1200	4.825	00:57:42						
1,040,203	29	1554	994,47	1200	4.655	00:52:58						
1,040,204	29	2011	992,04	1200	5.741	01:00:11						
1,050,101	29	2198	975,59	1200	5.801	01:04:05	10,501	95,37%	21,25	3:53:27	Morning	FRI
1,050,102	17	1672	983,66	1200	5.699	01:00:05						
1,050,103	14	957	1028,31	1200	4.468	00:48:59						
1,050,104	39	2121	1061,32	1200	5.283	01:00:18						
1,050,201	21	1216	842,96	1200	6.356	00:59:04	10,502	82,83%	22,01	3:47:18	Afternoon	
1,050,202	24	1827	833,2	1200	4.908	00:56:27						
1,050,203	22	1248	872,43	1200	5.441	00:55:37						
1,050,204	25	1448	1101,85	1200	5.301	00:56:10						
Total (38)	1054	70,973	37,911,76		269.619	42:30:01						

Table 11.
Collection circuits and sector results, showing the final numbers with routes and schedule of one vehicle during the week.

(a) (b) (c)

Figure 13.
Routes for sector 10,102 - (a) route for circuit 1,010,201, (b) route for circuit 1,010,202, and (c) route for circuit 1,010,203.

condominium (Sector: 010201), while the Vista Alegre, Marf II, and Vila do Sol condominiums (Sector: 010202) must be covered on Tuesdays as required by the association. In this way, operations are as close to Tuesday as possible. Friday had zone 0105 with the highest service load; however, it can be changed by moving the service of zone 0105 to Thursday or zone 0101 to Monday.

Alternative scales can be made, as a discharge time of 20 minutes was considered, and this can be reduced over time.

5. The effect of savings in domiciliary waste collection

Using as parameter the spreadsheets of costs publicly available in Brazilian municipality bids, there are mainly two centers of costs considered: staff and vehicle (or equipment). In the staff spreadsheets, the costs consider labor hour, unsanitary, extra time (day or night), working hours on holidays (day or night), vacation, food and transportation, termination, and social security according to the category's municipal syndicate rates. For the vehicle or the equipment spreadsheets, the costs consider maintenance and fixed costs, tires and tubes, lubrication and washing, average costs of vehicle consumption per km or per hour in operation while stopped, depreciation, annual license, insurance, and, finally, the percentage of the invested capital to have the vehicle if it is included. All these are composed in a final cost, which is approximately 60% consumed by the staff and 40% by the vehicle and/or equipment used. Depending on the value of the fossil fuel (diesel) used and distance traveled by the fleet, this difference can be lower, but it also suffers from syndicate annual intervention on the labor costs of collection.

Figure 14 presents the behavior of waste collection in downtown Petrópolis/RJ, where (a) presents the loading per day and (b) presents the distance traveled collecting in the sectors that are covered from Monday–Saturday at night with an 8.2-ton compactor and three garis per vehicle.

In sector routing, the major impact on savings is in the reduction of the number of sectors, meaning fewer vehicles used, and in the total distance traveled. In our evaluations in the field, we also found an expressive reduction in the staff for domiciliary waste collection. Although it was indicated in Plan A that it is better to use four men

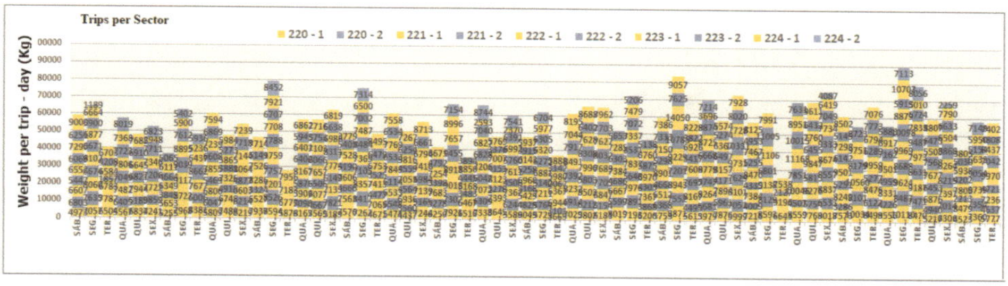

(a)

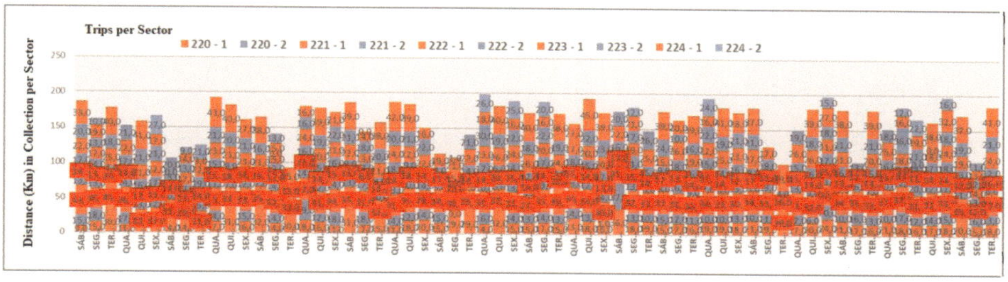

(b)

Figure 14.
Domiciliary waste collection in downtown Petrópolis/RJ. (a) Load per day – Downtown Petropolis/RJ. (b) Distance travelled per day – Downtown Petrópolis/RJ.

instead of three on daily-night collection in downtown Petrópolis/RJ, the global savings can be confirmed with our work of 7.52%. But, if the concessionaire wanted to use three men on Tue-Sat instead of four, it could bring a result in global costs savings of 14.04% (**Table 12**).

Table 12 presents an approximation of the global expenses with the vehicles and staff for the actual execution in comparison with Plan A and Plan B. These expenses depend on the costs according to the labor laws and the costs of maintenance of the vehicles, including expenses with fuel, tires, and so on. We did not include the costs of a vehicle, because the remained vehicle will be operated on Mondays in any case.

Considering the global costs (fixed and variable) of absorbing the sector-routing technology, as in **Table 12**, the results we obtained can be considered in the graphics exposed in **Figures 15** and **16**. The figures present three parts of bars corresponding to the costs of investing in the technology presented in this manuscript - project to revise sectors and routes of the city; for the amortization cost, the interest rate was set to

COSTS FOR PETROPOLIS/RJ (× R$ 1000,00)								
	Executed		**Plan A**		**Plan B**		**SAV A**	**SAV B**
Staff	R$ 132,13	59,50%	R$ 126,67	61,68%	R$ 112,19	58,77%	−4,13%	−15,09%
Vehicles	R$ 89,95	40,50%	R$ 78,70	38,32%	R$ 78,70	41,23%	−12,51%	−12,51%
	R$ 222,08		R$ 205,37		R$ 190,89		7,52%	14,04%

Table 12.
Global savings including all the real costs with staff and vehicles spences.

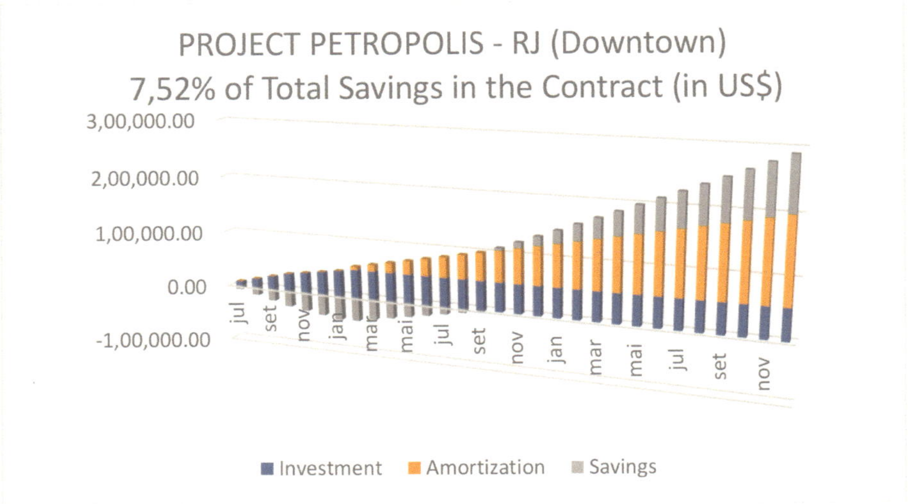

Figure 15.
Recovering investment when considering the team of collectors of plan a, returns start 15 months after project's end (may/2024).

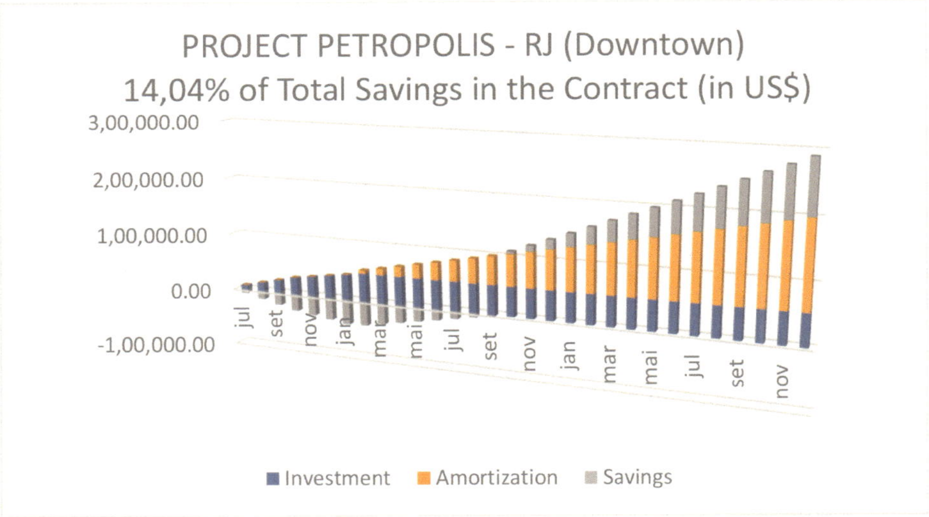

Figure 16.
Recovering investment when considering the team of collectors of plan B, returns start in Oct/2023, 9 months after project's end.

0.5% per month, and a 12% overprice in collection expenses from December to February, because of the rainy season and tourism overflow in downtown. For last the savings obtained with the results of the application.

The project returns the savings as the sectors and routes are implemented by the concessionary in the city; it starts returning as it starts to run. The breakeven point with the investment occurs in the 15th month from the beginning of the project for 7.52% of savings over spends in operational costs if Plan A is adopted (**Figure 15**). The breakeven point between savings and investment is reduced to the 9th month if the savings are of 14.04% (**Figure 16**).

6. Conclusion

The circuits generated and executed from October 2021 to January 2022 in downtown Petrópolis/RJ provided a reduction of over 17% in the total distance traveled by the fleet. It demonstrates that the plans generated by the framework can be executed with no differences between the planned and the executed results.

Regarding the cost of collector working hours, the average savings of 5% is relevant because the same work was performed with nearly the same amount of man-hours while covering 17% less distance, therefore increasing productivity. The savings with fixed and variable costs of over 17% were considerable, primarily because one driver, a crew (three men), and a vehicle were removed from the operation, eliminating their associated costs.

Another important aspect of the framework was the use of the CherryTrack® LIX Driver and CherryTrack® LIX Manager applications. The CherryTrack® LIX Driver, installed on a mobile device inside the load packer, assisted in navigation by indicating to the driver the sequence of roads he should follow to comply with the optimized route. Drivers were thought to have to "decorate" the route as the service execution relies on them. Embellish, CherryTrack® LIX broke this paradigm, which has dominated the medium for years. CherryTrack® LIX Manager allowed researchers, managers, and inspectors to monitor the service execution for each vehicle or the entire operation in real-time.

The project developed for RECICLEIROS in Bom Jesus dos Perdões/SP exceeded expectations. The new methodology can bring a better decision on recycling plant placement, and reasonable routes were designed and scheduled considering local constraints not fulfilled by the system but using manager reasoning.

OPILM SisRot® LIX, through its algorithms and software, reduced the operational costs of domiciliary waste collection in Petrópolis/RJ. The proposed SARP-based can be used for favorable tactical, operational, and strategic planning in waste management with success in other cities.

Acknowledgements

We would like to thank Força Ambiental and Recicleiros for their commitment while doing this work. We would also like to thank the Graphvs Ltda development team for facilitating this research.

Conflict of interest

The authors declare no conflict of interest.

Nomenclature

DSS	Decision Support System
SDSS	Spatial Decision Support System
OPILM	Optimized Planning and Integrated Logistics Management
GIS	Geographical Information System

GBMS	Graph Based Modeling System
ARP	Arc Routing Problem
CCP	Chinese Postman Problem
RPP	Rural Postman Problem
MRPP	Mixed Rural Postman Problem
CARP	Capacitated Arc Routing Problem
SARP	Sector Arc Routing Problem
ERP	Enterprise Resource Processing
WGIS	Web Geographical Information System
GPS	Global Positioning System
GPRS	General Packet Radio Service
RFID	Radio Frequency Identification
SQL	Structured Query Language
SisRot®LIX	Framework of SARP, GBMS and GIS systems
SisRot®FULL	Framework of VRP, GBMS and GIS systems
CherryTrack®LIX	Mobile Application to waste collection
VRP	Vehicle Routing Problems (Node Routing)
VPS	Virtual Private Server

A. Appendix

We show in more detail figures revealing more functionalities of CherryTrack®Lix with the apps Driver and Manager that was applied in the city of Petrópolis/RJ.

CherryTrack® Lix in operation.

We finalize with **Figures 17** and **18** showing the app CherryTrack®Lix in operation. In **Figure 17**, we have the best mounting of the smartphone in the vehicle cabin to help driver while in the trip. The app shows the next three blocks' movements while in collection and the next path between two distant collections, guiding the driver in the task of collection or just deadheading. In **Figure 18**, the manager version is shown; (a) shows to the supervisor the places where the collection must be done; (b) shows for a circuit of a selected sector the coverage of the circuit, while in red are the

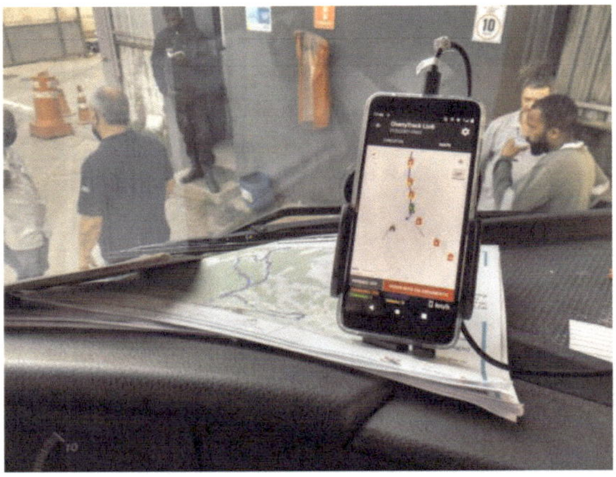

Figure 17.
Position of the CherryTrack® lix driver in the vehicle, including maps of the region of domiciliary waste collection.

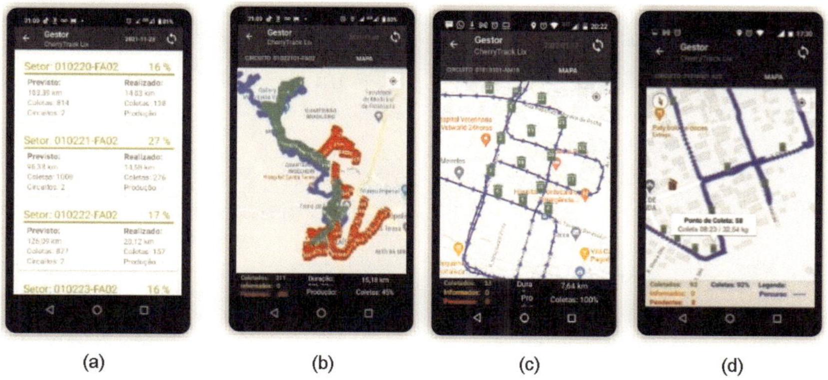

Figure 18.
New visualization of CherryTrack® lix manager, including maps of the region covered and left to be covered in the same trip of domiciliary waste collection.

non-collected street segment and in green are the collected segment; (c) the detailed movement performed by the driver in collection can be seen; and finally (d) reveals the load of collection obtained for each street segment in the path.

Author details

Marcos Negreiros[1]*, Augusto Wagner Palhano[2] and Eduardo Reis[3]

1 State University of Ceará, Campus do Itaperi, Fortaleza, Brazil

2 Graphvs Consultoria Ltda, Fortaleza, Brazil

3 Reis Consultoria Ltda, São Bernardo do Campo/SP, Brazil

*Address all correspondence to: marcos.negreiros@uece.br

References

[1] Graphvs & Barufi. Modelos e Planos de Otimização para a Integração do Sistema Logístico de RSU da Cidade de São Paulo", PMISP-2018,. São Paulo Parcerias/PMSP; 2018

[2] Mourão MC, Nunes AC, Prins C. Heuristic methods for the sectoring arc routing problem. European Journal of Operational Research. 2009;**196**(7): 856-868

[3] Caliper®. "TransCAD. Transportation planning software". Available from: https://www.caliper.com/ [Accessed: Ago/2022].

[4] Graphvs, "SisRot Lix". 2023. Available from: http://www.graphvs.com.br/?p=65 [Accessed: February, 2023].

[5] Corberán A, Prins C. Recent results on arc routing problems: An annotated bibliography. Networks. 2009;**56**(1):50-69. DOI: 10.1002/net.20347

[6] Simonetto EO, Borenstein D. A decision support system for the operational planning of solid waste collection. Waste Management. 2007;**27**: 1286-1297

[7] Golden B, Wong R. Capacitated arc routing problems. Networks. 1981;**11**: 305-315

[8] Ghiani G, Mourão C, Pinto L, Vigo D. Routing in waste collection applications, chapter 15. In: Corberán A, Laporte G, editors. MOS-SIAM Series on OptimizationArc Routing Methods and Applications. 2014a

[9] Ghiani G, Laganà D, Manni E, Musmanno R, Vigo D. Operations research in solid waste management: A survey of strategic and tactical issues. Computers & Operations Research. 2014b;**44**:22-32

[10] Han H, Ponce-Cueto E. Waste collection vehicle routing problem: Literature review. Traffic & Transportation. 2015;**27**(4):345-358

[11] Mourão MC, Pinto LS. An updated annotated bibliography on arc routing problems. Networks. 2017;**70**(3): 144-194

[12] Rodrigues AMM, Ferreira JS. Waste collection routing—Limited multiple landfills and heterogeneous Fleet. Networks. 2015;**65**(2):155-165. DOI: 10.1002/net.21597

[13] Boyaci B, Dang TH, Lechford AN. Fast upper and lower bounds for a large scale real world arc routing problem. Networks. 2023;**81**(1):107-124. DOI: 10.1002/net.22120

[14] Janela J, Mourão MC, Pinto LS. Arc routing with trip-balancing and attractiveness measures — A waste collection case study. Computers & Operations Research. 2022;**147**:105934

[15] Rodrigues AMM. Sectores e Rotas na Recolha de Resíduos Sólidos Urbanos [thesis]. Portuguese: University of Porto; 2014

[16] Negreiros M, Palhano A, Batista P. The heterogeneous sector routing problem, Workshop in Arc Routing Problems (WARP II). Lisbon, Portugal: ISEG; 2016

[17] Batista P, Negreiros M, Muritiba A, Palhano A. New framework of metaheuristics for the capacitated Centred clustering problem. Annals of xI Metaheuristics International Conference (MIC). 2015

[18] Corberán A, Martí R, Romero A. Heuristics for the mixed rural postman problem. Computers & Operations Research. 2000;**27**:183-203

[19] Ghiani G, Guerriero E, Manni A, Manni E, Potenza A. Simultaneous personnel and vehicle shift scheduling in the waste management sector. Waste Management. 2013;**33**:1589-1594. DOI: 10.1016/j.wasman.2013.04.001

[20] Ghiani G, Manni M, Manni E, Toraldo M. The impact of an efficient collection sites location on the zoning phase in municipal solid waste management. Waste Management. 2014c;**34**:1949-1956

[21] Wøhlk S, Laporte G. A districting-based heuristic for the coordinated capacitated arc routing problem. Computers & Operations Research. 2019;**111**:271-284

[22] Cortinhal MJ, Mourão MC, Nunes AC. Local search heuristics for sectoring routing in a household waste collection. European Journal of Operational Research. 2016;**2016**:1-12. DOI: 10.1016/j.ejor.2016.04.013

[23] Klashner R, Sabet S. A DSS design model for complex problems: lessons from mission critical infraestructure. Decision Support Systems. 2007;**43**:990-1013

[24] Negreiros M, Palhano A, Rodrigues JA, Chaves B, Albuquerque I. Routing, tracking and drivers assistance in wholesale – SisRot®, cherry/Zeus-track®. IFAC Proceedings. 2013;**46**(24):523-526. DOI:10.3182/20130911-3-BR-3021.00123

[25] Negreiros M, Xavier AFS, Lima JWO, Xavier AE, Maculan N, Michelon P. Integração de sistemas computacionais e modelos logísticos de otimização para prevenção e combate à dengue. Pesquisa Operacional. 2008;**28**(1):1-27

[26] RouteSmart™ Technologies. Available from: https://www.routesmart.com/public-works

[27] Environmental Expert. "Public Works Solutions LLC (PWS) Software". Available from: https://www.environmental-expert.com/companies/public-works-solutions-llc-pws-25071/software [Accessed: Ago/2022].

[28] Sensoneo. "Global leader in smart waste solutions". Available from: https://sensoneo.com/route-planning/ [Accessed: Ago/2022].

[29] EasyRoute™. "A routeware global company". Available from: https://www.goeasyroute.com/ [Accessed: Ago/2022].

[30] IBGE. Instituto Brasileiro de Geografia e Estatística. 2018. Available from: https://cidades.ibge.gov.br/brasil/ [Accessed: December, 2018].

[31] IBGE. Instituto Brasileiro de Geografia e Estatística. 2020. Available from: https://cidades.ibge.gov.br/brasil/ [Accessed: December, 2020].

[32] IBGE, Instituto Brasileiro de Geografia e Estatística. 2017. Available from: https://cidades.ibge.gov.br/brasil/ [Accessed: October, 2017].

[33] Recicleiros. 2023. Available from: https://recicleiros.org.br [Accessed: February, 23].

[34] IBGE, Instituto Brasileiro de Geografia e Estatística. 2010. Available from: https://cidades.ibge.gov.br/brasil/ [Accessed Jun/2011]

[35] Mcbride R. Controlling left and u-turns in the routing of refuse

collection vehicles. Computers & Operations Research. 1982;**9**(2):145-152. DOI: 10.1016/0305-0548(82)90013-2

[36] Bodin L, Fagin G, Welebny R, Greenberg J. The design of a computerized sanitation vehicle routing and scheduling system for the town of oyster bay, New York. Computers & Operations Research. 1989;**16**(1):45-54. DOI: 10.1016/0305-0548(89)90051-8

[37] Negreiros M, Palhano A. Line Graph Transformation to the Euler Tour with Movement Prohibition Problem. Melbourne, Australia: Annals of IFORS; 2011

Chapter 3

Perspective Chapter: Environmental-Friendly Agro Waste Management

Manabendra Patra and Duryodhan Sahu

Abstract

Abundant amount of agro wastes is produced day by day globally to manage the escalating needs of billons of human population. The agro wastes are produced from various sources mainly crops left out, agro industries, aquaculture, and livestock. The major ingredient of agro wastes are of cellulose, lignin, hemicelluloses, etc. Conventionally, most of the crops left out were used for composting, animal fodder, domestic fuel, etc. Due to modernization technology in agriculture sector, people from Third World countries prefer cost-effective methods such as combustion process. Improper management of agro waste generated in the process has been contributing toward escalating air, soil, and water pollution. A proper environmental-friendly management of agro waste is the need of the time for sustainability, food, and health security of human. Lignin and hemicellulose can be used for generation of biofuels and biofertilizer. Cellulose can be sustainably used for the production of nanosilica, biodegradable polymer, paper, pulp, etc. This chapter emphasizes sustainable agro waste management without affecting the environment at lower cost in timely manner. In particular, the agro waste biomass could be used as a source of value-added bio-product, which has wide applications and impacts the bio-economy without hampering the climatic change issue.

Keywords: agro waste, composting, activated carbon, nanocellulose, biofuels, ethanol

1. Introduction

Agro waste is called undesirable material that is generated from agriculture farming practices and includes crop residues, leaf litter, livestock waste, sawdust, weeds, and forest waste. Agro waste is misled or discarded in most part of this earth due to either unawareness or proper route to transfer and its utilization. These agro wastes can create a great constraint, if proper measures will not be taken off for its proper discard, as it can lead to a clearly impact on the environment issue. In addition to the hazards of burning and land filling, the synthetic chemicals adopted during farming and agriculture are able to instigate pollution if these wastes wind up in the undesirable places. Agriculture sector is the major backbone of the developing nations, and it is one of the greatest contributors to the GDP. Millions of people are adopting agriculture as their primary occupation in this earth. With the ever-increasing of the world population, there is an escalation in the demand for food and food product supplies, so nowadays many people make use of modern agriculture to encounter the need. Modern agriculture uses the up-to-date cultivation technique along with synthetic fertilizers. Urban people are also adopting garden farming using modern methods. The demand for animal products such as milk products, poultry, and meat is also high, and producers have been developing new strategy to enhance the productivity and lower the unit cost of production. Chemicals such as fossil fuels, inorganic fertilizers, and organic pesticides improved genetics of production species and are enhancing the increase in the generation. The agro waste and its processing are a universal issue, as its major part is going for burning or to be buried in soil, which is responsible for pollution of air, water, and soil, a general loss of aesthesis. The degradation of water quality can affect adjoining water bodies and groundwater both on-site and off-site. Such kind of degradation in the water quality reduces the ability of water resource to support aquatic life and water consumption for humans and animals. Unplanned burying of agro waste leads to greenhouse emission and a major concern for climatic change. Conventionally, a large volume of agro waste is utilized as animal fodder, domestic combustion fuel, composting, roof thatching, etc. The management of agro waste and the reformation of its transition into a fit for use product through the utilization of biotechnology in agriculture are getting a lot of recognition nowadays [1]. Solid state fermentation can be considered as the better process for transition of agro waste into usable bioproducts. Different agro wastes such as wheat straw [2], barley straw [3], cotton stalks [4], sunflower stacks [5], etc., from abundant agriculture goods, as well as significant horticulture wastes such as apple [6], mango [7], orange peels [8], and potato [9] were used to create beneficial products in this review. Agro wastes can be used for contributing to guaranteeing resource efficiency, sustainable production and consumption, and the reduction of negative environmental impact on adopting recent and superior scientific methods in disposal of it [10]. The main objective of this chapter is to emphasize on the use of agro waste as a source for generating ecofriendly material such as (i) compost to enhance soil productivity, biodiversity, and sustainable environment; (ii) activated carbon, which can be used as adsorbent for removal of heavy metals from drinking and industrial wastewater; (iii) nanocellulose, which is widely applicable as membrane in water purification; (iv) the agro waste containing lignocellulosic residues can be used to produce different bioproducts including biofuel, biofertilizer, bioplastic, organic acid, etc. Furthermore, these value-added products can enhance the bioeconomy without affecting the environment.

2. Methods of agro waste utilization and management

2.1 Composting

Composting is one of the old-age methods for transforming agro wastes into hygienic, stabilized, and non-polluting materials, thus retrieving the beneficial nutrients and enhancing soil fertility [11]. Therefore, composting is commonly adopted as a biological treatment method for agro wastes, but enhancing compost maturity is essential for the safe use of composting products [12]. The agro waste can be converted into valuable compost on utilization of proper role of certain microorganism. The generated product obtained by microbial action has a lot of superiority in agriculture as it helps in enhancing productivity, better soil biodiversity, and sustainable environment. Thus, composting can be one of the better options for the processing of the large volume of agro wastes generated worldwide [13]. Gusmawartati et al. [14] have studied the quality of compost generated from taking various combinations of agro wastes such as cassava peel, empty fruit bunches of oil palm, banana skin, and rice straw. Mastouri et al. [15] have studied the growth of lettuce using compost obtained from a mixture of tree bark wastes obtained from orchid, aldar, horn beech, oak, hard wood tree, etc. Pergola et al. [16] had emphasized on the restoration of soil organic matter function in agricultural soil with various agricultural additives of reconverted waste biomasses. Aslam et al. [17] have studied vermin composting of rice straw, wheat straw, and cow dung by Eisenia fetida on-farm management of nutrients such as NPK, beneficial humus, soil microbes, phosphate-solubilizing bacteria, actinomycets, micronutrients, nitrogen fixing, growth hormones such as auxins, etc. Yu et al. [18] had taken multiple combinations of agro waste (from mushroom industry) compost and biofertilizer for enhancing yield and higher sustainability of a pepper crop. Trillas et al. [19] have studied the reduction of solani diseases in cucumber seedlings by application of compost obtained from agro waste such as olive marc, grape marc, spent mushroom, and cork. Particularly, Trichoderma Asperellum reduces the relevance of solani pathogen in the soil on amending at 103 cfu/ml. Karak et al. [20] had investigated the maturity of compost obtained with various ratios of agro wastes, such as wheat straw, rice straw, mustard stover, and potato plant obtained in both the presence and absence of fish pond bottom sediment. Gea et al. [21] had studied on controlling dry bubble diseases in mushroom farming caused by Verticillium fungicola. They had used various agro-based waste composts generated from a mixture of olive oil husk, used mushroom substrate, cotton grit thrashed along with compost from grape marc compost, rice husk, and cork compost.

2.2 Activated carbon

The activated carbon, a low-cost and high-quality material, can be utilized for adsorption purpose in various applications. Generally, the activated carbon is prepared by burning lignite, coal, wood, etc., in pyrolysis over 600–900°C. Nowadays, a lot of emphasis has been given to lignocellulosic biomass, readily available in agriculture sector as waste, for generation of the activated carbon [22]. The activated carbon generated from agro-based waste has its own advantage due to its low cost and ubiquitous availability [23]. During the last few decades, there was growing research interest on the utilization of alternative origin of waste materials from industry and agriculture for activated carbon production [24–27]. A sizable numbers of reports have been published on generation of activated carbon from agro waste such as palm shell [28, 29], coconut shell [30, 31], corn cob [32], olive stones [33] and walnut shell [34], coir pith [35], rice bran [36], chickpea husks [37], oil palm shell [38], etc. Ioannidou et al. [23] have used the agricultural residues such as soya stalks, corn cobs, rapeseed stalks, and olive kernels as precursors for the generation of activated carbon. The pyrolysis were done in two stages: (a) the pyrolysis had been carried out over 800°C for about 1 h under nitrogen atmosphere (15 ml/min) along with heating rate at 27°C/min for the sake of producing char, (b) the physical activation of char was then carried out over 800°C for about ½ h under the flow of steam (40 g/min) at pressure of ½ bar. The obtained activated charcoals were subjected to study of removal of Bromopropylate, common pesticides in fruits crop, from water. Tay et al. [39] have isolated activate carbon on pyrolysis of soybean oil cake by chemical activation with potassium hydroxide and potassium carbonate at different temperatures of 600 and 800°C. Potassium carbonate was found to be more effectual as compared with potassium hydroxide under similar conditions.

The maximum surface area of activated carbon obtained with potassium carbonate at 800°C is found to be 1352.86 m^2/g, which is in accordance with the range of commercial activated carbons. Anne A. Nunes et al. [40] derived activated charcoal from defective coffee press cake by heating it under nitrogen atmosphere at 600/800°C for elimination of methylene blue from water up to 99% removal. The maximum adsorption capacity obtained for the coffee cake activated carbon/methylene blue system was observed to be 14.9 mg/g. The equilibrium data fitted favorable into Freundlich model as compared with others. Rice is one of the widely grown crops in the world, generating a large volume of waste. That has to deal with proper management due to short duration in between two crops. During last few decades, people have reported on generation of activated carbon from rice straw [41]. The utmost value of carbofuran adsorption capacity was observed to be 26.52 mg/g. Chang et al. [42] had studied elimination of bisphenol-A from water by using activated carbon obtained by with the help of chemical (potassium hydroxide) treatment of rice husk. At pH 2.5, the maximum adsorption capacity of bisphenol-A was found to be 181.191 mg/g. The experimental values perfectly fitted the Langmuir model for equilibrium data. It was found to be more inclination toward pseudo second order as compared with that pseudo first order. Isoda et al. [43] reported the generation of activated carbon from rice husk with more surface area, of about 1500 m^2/g and high mesopore volume of about 1.22 cm^3/g using chemical (zinc chloride) treatment over an activation temperature of 600°C without carbonization and using sodium hydroxide as chemical activating agent, with carbonization. Köseoğlu et al. [44] had studied the generation of activated carbon from orange peels using potassium carbonate and zinc chloride as chemical reagents for the purpose. The surface area of the activated carbon was observed to be 9–1352 m^2/g for potassium carbonate and that for zinc chloride 804–1215 m^2/g. Potassium carbonate was observed to have much potential as compared with zinc chloride as a chemical activating reagent in light of high surface area, development of porosity, and surface analysis of the activated carbon. Mahamad et al. [45] had studied the generation of activated carbon from solid pine apple waste mass such as leaves, stem, and crown using zinc chloride as chemical reagent at 500°C for 1 h. It can be deduced that the activated carbon obtained by a 1:1 ratio has the better removal of dye capacity, which can be attributed to its high surface area (914.67 m^2/g) and dye adsorption capacity (288.34 m^2/g). The Langmuir adsorption isotherm model is perfectly suited to the obtained adsorption equilibrium data with R^2 of 0.969. The maximum uptake of methylene blue with obtained activated carbon was observed to be 288.34 m^2/g. Baysal et al. [46] had prepared activated carbon from sunflower piths using NaOH and KOH as chemical reagents. The activated carbon prepared has high surface area of about 2690 and 2090 m^2/g. Activated carbon obtained from mahogany fruit shell successfully used for about 99.7% uptake of lead ion from wastewater [47] . Xue et al. [48] have used Angelica keiskei as a source for generation of activated carbon, which could be utilized as efficient adsorbent of organic dyes from wastewater. As shown in **Table 1**, activated carbon made up of various agro wastes with different temperatures of activations used as adsorbent for the removal of different impurities from wastewater. Furthermore, the maximum capacities of removal are complying well with the disposal standard of wastewater.

2.3 Nanocellulose from agro waste

The nanocrystalline cellulose can be generally isolated from different subsequent chemical process: starting with bleaching and alkali treatment succeeded by acid

Adsorbent source	Adsorbate	Activation temperature (°C)	Capacity	References
Soya stalks, Corn cobs, Rapeseed Stalks and Olive kernels	Bromopropylate (isopropyl 4,4′-dibromobenzilate)	800	0.0948 mg/g	Ioannidou et al. [23]
Coffee Press Cake	Methylene blue	600/800	14.9 mg/g	Nunes et al. [40]
Rice Straw	Carbofuran	850	296.52 mg/g	Chang et al. [41]
Rice Straw	Bisphenol-A	850	181.19 mg/g	Chang et al. [42]
Orange Peel	Iodine Methylene Blue	500–1000	1564 mg/g 150 mg/g	Köseoğlu et al. [44]
Pineapple Waste	Methylene Blue	500	288.34 mg/g	Mahamad et al. [45]
Sunflower Piths	Methylene Blue	500	965.349 mg/g	Baysal et al. [46]
Mahogany Fruit Shell	Pb(II)		322.28 mg/g	Patil et al. [47]
Ashitaba waste	Methylene Blue	900	491.56 mg/g	Xue et al. [48]

Table 1.
Activated carbon generated from various agro wastes.

hydrolysis of the natural fibers. The nanocellulose isolated from various sources of agro waste is becoming an attractive research avenue for its multifaceted utilization [49]. Nowadays, a lot attention has been given to generation of nanocellulose from various ago wastes such as olive tree pruning [50], pine cones [51], pineapple leaf [52], rice husk [53], sisal fiber [54], sorghum stalk [55], sunflower stalks [56], etc. Ferreira et al. had successfully isolated cellulose nanocrystals from sugarcane bagasse, on hydrolysis by sulfuric acid, which had very good hydrophilic properties with a high crystallinity. Adipic acid was used for surface modification of nanocrystal for suppressing the crystal dimension by elimination of amorphous region [57]. Johar et al. reported on the isolation of nanocellulose fibers from rice husk. They adopted alkali (NaOH) and bleaching ($NaCl_2O$) treatment followed by acid (H_2SO_4). They observed a remarkable enhancement in crystallinity of the obtained nanocellulose [53]. Lu and Hsieh et al. had extracted an unblended form of nanocellulose from rice straw with about yield of 36%. The acid hydrolysis for about ½ h resulted in nanocellulose of size of 270 nm length and 5.95 nm diameter, whereas acid hydrolysis for 45 min resulted in nanocellulose of size of 117 nm length and 5.06 nm diameter [58]. do Nascimento et al. had successfully extracted cellulose nanocrystals from coconut fiber [59]. de Carvalho Mendes et al. isolated crystalline nanocellulose from various agro wastes such as garlic skin, palm oil, sesame, and rice husks [60] . Walnut shell (*Juglans regia* L.) was utilized as the raw material for the production of purified cellulose [61]. The lignin and hemicellulose present in walnut shell had been perfectly eliminated by sodium hydroxide treatment and followed by bleaching with equal amounts of 1.7 wt.% sodium chlorite and acetate buffer solution, which leads to the enhancement of cellulose content up to 89%. Sijabat et al. had isolated nanocellulose from waste media of Kepok bananas (*Musa paradisiaca* L.) applying Gluconacetobacter xylinusbacteria in the fermentation procedure in utilization for membrane applications in water filter [62].

2.4 Biofuels

The second-generation biofuels, commonly prepared from inedible crops, woody crops or lignocellulosic biomass, agro waste, or unwanted plant, are potent reply to the food versus fuel feud as they utilize leftover portion of agro waste. Inedible feedstock is commonly used for the second-generation biofuels, i.e., jatropha, grasses, wastes vegetable oil, wood chips, etc. Alcohol generation from rapid growth plants could be produced by enzymatic activities to isolate out the sugars from lignin fibers of the biomass. Syngas, a mixture of hydrogen and carbon monoxide, can be synthesized on thermochemical treatment of biomass. Hydrogen thus prepared can be used as fuel, and other hydrocarbons can be used as add-on to the gasoline [63] . Recently, most of the gasoline available is blended with certain percentage of ethanol to reduce carbon footprint. The effective conversion of cellulose into ethanol has got major prospective due to the ubiquitous obtainability, plentitude, and comparable inexpensive cellulosic plant materials. The banana residue includes banana fruit (pulp and peels) and lignocellulosic biomass can be a potential source for biofuels [64]. Srivastava et al. [65] had successfully utilized Saccharomyces cerevisiae for generation of bioethanol out of rice husk up to yield of 250 mg/g dry biomass after 6 days of fermentation. Singh et al. [66] had enzymatically hydrolyzed the pretreated rice husk with alkali under microwave condition for the generation of biofuel. They have successfully utilized Scheffersomyces stipites and S. cerevisiae yeast for the fermentation. The ethanol production with S. cerevisiae was to be 0.3–0.39 g/g; with Scheffersomyces stipites, waste 0.24–0.35 g/g, respectively. Chukwuma et al. [67] had adopted fermentation process of rice husk using Aspergillus fumigatus, Aspergillus niger, and Saccharomyces cerevisae for the generation of biofuel. On fermentation with Aspergillus fumigatus, treating rice husks shows the at most cellulose of 45 ± 3.31%, hemicelluloses of 31 ± 3.00%, reducing sugar of 2.60 ± 0.30%, carbohydrate of 19.52 ± 10.05%, and non-reducing sugar of 16.92 ± 9.75% producing ethanol yield of 6.60 ± 0.48% with palm wine yeast, while 5.60 ± 0.42% yield was with bakers. Slow pyrolysis activity by thermogravimetric analysis had been investigated to estimate and compare the effective utilization agro waste such as corncob, rice husk, wood chips, wheat straw, bagasse, etc., for biofuel conversion [68]. The corncob was observed to deteriorate with an enhanced rate over lower temperature. On the whole, the activation energy was observed to be enhanced at the reduced temperature range (250–400°C), and that was reduced in the enhanced temperature range (450–600°C). The corncob had been observed to be a suitable contender out of the rest of the wastes for pyrolysis with activation energy of 29.71 and 4.23 kJ/mol in reduced and enhanced temperature range, respectively. Buenrostro-Figueroa et al. [69] had used *Kluyveromyces marxianus* for fermentation of mango fruit for ethanol generation. *K. marxianus* in Tommy Atkins mango juice exhibits encourage finding over Haden mango juice. The finding showss that of 4 g/l/day, a yield of up to about 49% of ethanol and a process efficiency of about 80%. Mihajlovski et al. [70] had utilized Streptomyces fulvissimus for fermentation of lignocellulosic waste such as wheat bran, barley bran, and rye bran, to obtain alcohol. Rye bran observed to be one of the most perfect waste substrates that can be used for bioconversion. Najafi et al. [71] had reported enzymatic potential of the bacterial strain S. fulvissimus during the hydrolysis of lignocellulosic agro waste such as rice, wheat, sugar cane, barley, and corn for the generation of ethanol. Pistachio waste such as pruning trees, green (soft) shell, and hard shell could be transformed into beneficial fuel using the fermentation processes, anaerobic digestions, and thermochemical degradation (i.e., pyrolysis) methods for production of biofuels [72]. Yuliansyah et al. [73] had evaluated the feasibility of upgrading oil palm fronds and trunks for their decomposition behavior over hydrothermal treatment to generate solid biofuels. The rice straw biomass is constituted of different variety of biopolymers, mainly cellulose, hemicellulose, and lignin. Through the hydrolysis of cellulose and hemicellulose, monomeric sugars are liberated that can be converted into ethanol by fermentation as an alternative to biogas by anaerobic digestion. Laobussararak et al. had utilized the bacterium *Zymomonas mobilis* and distillery yeast, *S. cerevisiae* and a co-culture of *Z. mobilis* and *S. cerevisiae* for fermentation of rice straw waste for production of ethanol [74]. The rice straw had been treated with 2% sodium hydroxide solution, then followed by enzymatic hydrolysis making use of cellulase prior to the fermentation. It was found to be that 2% NaOH pretreatment is perfectly suitable for the rice straw waste as a type of pretreatment context able to generate the high cellulosic content about of 88.96% and diminishing sugar content of 9.18 g/l. The distillery yeast was found to be a befitting microorganism for the generation of ethanol out of the rice straw, as ethanol yield on enzymatic hydrolysis found to be 15.94–19.73%, 20.48–35.70%, and 21.56–29.89% for the bacterium, yeast, and co-culture, respectively. Kumar et al. [75] had comprehensively investigated over green solvent-pretreated rice straw and cellobiose fermenting yeast strain Clavispora for production of cellulosic ethanol. Green solvent (cholinechloride/glycerol) treated rice straw leads to maximum reduction of sugars about 226.7 g/l with a saccharified capability of about 87.1% at 20% solids loading and 12FPU cellicctec2. The generation of ethanol yield of 36.7 g/l was found out of 8% of glucose within 36 h with a conversion capability up to 90.1%. Sasaki et al. [76] had successfully studied that the perfluoropolymer membrane has been suitable used in vapor permeation to isolate aqueous ethanol from combined product obtained out of rice straw with recombinant S. cerevisiae. Kluyveromyces sp. is explored as thermophilic ethanologen, which effectually makes use of hexose for the fermentation of ethanol at high temperature (45–50°C) [77]. The rice straw waste had been hydrolyzed at temperature of 140°C along with dilute H_2SO_4 of 0.6%v/v over 90 min for utmost retrieval of pentose monomer yield of 12.52 g/100 g. Using commercial cellulose, saccharification efficiency was observed to be 79 ± 0.05% with acid-hydrolyzed biomass. The fermentation of saccharified broth utilizing thermophilic yeast Kluyveromyces sp. with cell recycle produced ethanol with an overall yield and productivity of 93.5 ± 0.05% and 0.90 ± 0.2 g/l/h, respectively, and with a negligible amount residual sugar found in fermentation broth. Assis Castro et al. [78] had investigated multiple approaches of saccharification as well as fermentation utilizing rice straw waste that is pretreated with dilute acid for ethanol production using thermo-tolerant yeast Kluyveromyces marxianus. On concurrently saccharification and fermentation, in the absence of type of any pre-hydrolysis, it was observed to be as the utmost perfect condition owing to the enhanced ethanol generation (1.4 g/l. h), about two times more in contrast to the alternate approach. Mahajan et al. [79] had accessed glycosyl hydrolases produced by different thermophilic fungal strains for the saccharification of alkali as well as biologically (Trametes hirusita/Myrothecium roridum) treated Parthenium hysterophorus (carrot grass) as well as rice straw waste. The integrative examination of hydrolysates observed clear-cut outline of hexose, pentose, and oligomeric sugars. Malbranchea cinnamomea was utmost orderly origin of glycosyl hydrolases producing 283.8, 35.9, 129.6, 27,193, 4.66, 7.26 (units/gds) of endoglucanase, cellobiohydrolase, b-glucosidase, xylanase, a-arabinofuranosidase, and b-xylosidase, respectively. The fermentation of outcome hydrolysates having glucose/xylose was competently yield of ethanol by *S. cerevisiae* due to the presence of xylose isomerase (0.8 units/gds) activity in culture extract of *M. cinnamomea* resulting in generation of 16.5 and 15.0 g/l of ethanol from alkali-treated rice straw and carrot grass, respectively. Sasaki et al. [80] had utilized a xylose-fermenting *S. cerevisiae* strain in ethanol fermentation activities for accessing in perfect usage of hemicellulose generate from rice straw waste. The xylose fermenting recombinant *S. cerevisiae* helps in generating bioethanol yield of about 34.5 ± 2.2 g/l. Momayez et al. [81] had investigated utilizing the liquid anaerobic digester, the biogas liquid waste, for the pretreatment process of the rice straw at different ambience. The rice straw had been pretreated at varying temperature 130–190°C for ½/1 h duration and put through to enzymatic hydrolysis, simultaneous saccharification and fermentation, dry anaerobic digestion, and liquid anaerobic digestion. The hydrolysis is enhanced by 100%, while the yield of ethanol enhanced by 125% on treating the rice straw waste at temperature of 190°C over 1 hour. There was also enhancement of yield of methane in 24 and 26% on using pretreatment process of rice straw through liquid anaerobic digestion and dry anaerobic digestion. Molaverdia et al. [82] had utilized Mucor indicus fungus for fermentation of rice, which was pretreated with 0.5 M Na_2CO_3 solution over 3–10 h to enable improving the efficiency of ethanol production. The maximum ethanol yield of about 99.4 g/l generated from the pretreated rice straw waste on simultaneous saccharification and fermentation for about 10 h. Whereas the ethanol yield of about 66.3 g/l was generated on moderate enzymes loading for fermentation about 12 h. Lü et al. [83] studied improving ethanol generation by the pretreating the rice straw waste with the microwave-assisted $FeCl_3$ solution followed by applying simultaneous saccharification and fermentation using *S. cerevisiae* and Pichia stipites. The concentration of ethanol is about 5.51 g/l on fermentation. Trametes hirsute, a white rot fungus, was competed of directly fermenting starch, wheat bran, and rice straw, for generating ethanol in the absence of acid or enzymatic hydrolysis [84]. *T. hirsuta* appeared comparable xylose consumption and ethanol production with a yield of 0.44 g/g. On growing the fungus in a medium containing 20 g/l starch, wheat bran, or rice straw, it was observed that ethanol yield of 9.1, 4.3, and 3.0 g/l, respectively. Karimi et al. [85] had studied for ethanol production out-of rice straw, pretreated with dilute acid, with the yields of 74, 68, and 61% using Rhizopus oryzae, Mucor indicus, and *S. cerevisiae*, respectively. The yield of ethanol was found to be 74 and 68% while using *R. oryzae* and *M. indicus*, respectively. Kaur et al [86] had reported low-cost process involving solid state fermentation of rice straw producing high titers of cellulases and hemicellulases for hydrolysis of alkali pretreated rice straw leading to ethanol yield of 15.6 g/l. Sharma et al. [87] had utilized the charred wood ash from Acacia nilotica as the diversified base catalyst. The wood ash catalyst that charred at 800°C shows an improved catalytic effect owing to its enhanced surface area, which leads to produce a 98.7% biodiesel transformation. It was also observed that the wood ash catalyst was found to be steady for the reaction of jatropha oil without any leaching of catalyst material. Uprety et al. [88] had investigated synthesis of biodiesel from palm oil utilization of ash from Birch bark as a diversified catalyst. It was observed that the biodiesel yield of 69.70% in the effective context like a catalyst load of 3 wt.% while a methanol to oil molar ratio of 12:1 at temperature of 60°C within 3 h. Betiku et al. [89] reported synthesis of biodiesel from *Azadirachta indica* oil transesterification utilizing a catalyst obtained on calcination of cocoa pod husk ash. It was observed that the cocoa pod husk ash catalyst charred over temperature of 700°C producing good interest owing to its high potassium content of about 59.2% and formation of the microstructure. While taking catalyst amount of 0.65 wt.% and a methanol: oil ratio of 0.73 (v/v) at temperature of 65°C within 57 min, the yield of biodiesel transformation was found to be 99.3%. Vadery et al. [90] had investigated generation of biodiesel from jatropha oil by utilizing a catalyst on calcination of coconut husk. While taking catalyst amount of 7 wt.% and a methanol: oil ratio of 0.73 (v/v) at temperature of 45°C within 45 min, the yield of biodiesel transformation was found to be 99.86%. The present demand of the time is to find out the perfect catalyst that can be both eco-friendly and economical aside from showing excellent catalytic activity for synthesis of biodiesel. Nowadays, a lot of attention is given on the designing of perfect green catalysts derived out of agro-based waste for the trans-esterification of vegetable oils, banana stem ash [91, 92], and banana peel ash [93–97], are few examples of the catalysts, obtained from agro-based waste, which had been perfectly made use of as basic catalysts for the generation of biodiesel.

3. Removal of heavy metal

A significant deal of interest has been focused in the research for the removal of heavy metals from industrial effluent using agricultural by-products as bio-adsorbents. The use of agro waste in bioremediation of heavy metal ions, i.e., biosorption utilizes inactive (nonliving) microbial biomass to bind and aggregates heavy metals from waste water by physicochemical pathways (mainly chelation and adsorption) of uptake [98]. Agro waste such as hazelnut shell, rice husk, pecan shells, jackfruit, maize cob, or husk can be used as bioadsorbent for heavy metal removal after chemical modification or conversion of these agro wastes into activated carbon.. Orange peel was employed for Ni(II) removal from simulated wastewater and was found maximum metal removal occurred at pH 6.0 [99]. Coconut shell charcoal (CSC) modified with oxidizing agents and/or chitosan was used for Cr(VI) removal was investigated well by Babel and Kurniawan [100]. Further, Cu(II) and Zn(II) were removed from real wastewater using pecan-shells-activated carbon [101] and potato peels charcoal [102]. The Cr(VI) removal from an aqueous solution by rice-husk-activated carbon has been studied extensively [103]. It was found that the maximum metal removal by rice husk took place at pH 2.0. Rice husk, containing cellulose, lignin, carbohydrate, and silica, was investigated for Cr(VI) removal from simulated solution [104]. To enhance its metal removal, the adsorbent was modified with ethylenediamine. The maximum Cr(VI) adsorption of 23.4 mg/g was reported to take place at pH 2. Other types of biosorbents, such as the biomass of marine dried green alga (biological materials) [21–25], were investigated for uptake of some heavy metals from aqueous solution. Some of the used alga wastes were *Spirogyra* species [105], Ecklonia maxima [106], *Ulva lactuca* [107], *Oedogonium* sp. and *Nostoc* sp. [108], and brown alga *Fucs serratus* [109]. On the whole, an acidic pH ranging 2–6 is effective for metal removal by adsorbents from biological wastes. The mechanism of uptaking heavy metal ions can take place by metabolism-independent metal binding to the cell walls and external surfaces [110]. This involves adsorption processes such as ionic, chemical, and physical adsorption. A variety of ligands located on the fungal walls are known to be involved in metal chelation. These include carboxyl, amine, hydroxyl, phosphate, and sulfhydryl groups. Metal ions could be adsorbed by complexing with negatively charged reaction sites on the cell surface shows the adsorption capacities of different biosorbents. Several studies have demonstrated the ability of rice husk to remove heavy metals from water sources. A study of the removal efficiencies of nine different heavy metals using rice husk observed maximum adsorption capacities ranging from 5.5 to 58.1 mg/g, with the values increasing in the following order: Ni(II) < Zn(II) < Cd(II) < Mn(II) < Co(II) < Cu(II) < Hg(II) < Pb(II) [111]. The rice straw and rice bran have been shown to remove Cu(II) with maximum adsorption capacities of 18.4 and 21.0 mg/g, respectively [112]. In a study on the use of rice husks for the adsorption of Cr(VI), significant removal (>95%) occurred in the case of low pH (<3.0), primarily due to the speciation of the Cr(VI) ions [113]. Bansal et al. [114] evaluated the removal of Cr(VI) using rice husk and achieved a maximum adsorption capacity of 8.5 mg/g; they also found that treating rice husk with formaldehyde enhanced removal by approximately 23%. Another study used phosphate-treated rice husk to evaluate the removal of Cd(II) from wastewater and achieved a high maximum adsorption capacity (103 mg/g at 20°C) [115]. Residuals from peanuts were also found to be an effective adsorbent for the removal of heavy metals. A maximum adsorption capacity of 39 mg/g was achieved for the removal of Pb(II) using peanut shells; significant removal was observed at various temperatures and pH conditions [116]. Peanut shells were also shown to removal of Cr(VI) at low pH values, achieving a maximum adsorption capacity of 4.3 mg/g [117]. Moreover, researchers achieved effective removal of Cr(III) and Cu(II) using peanut shells with maximum adsorption capacities of 27.9 and 25.4 mg/g, respectively [118]. Researchers also observed significant heavy metal removal with peanut husks, achieving maximum adsorption capacities of 7.7, 10.2, and 29.1 mg/g for Cr(III), Cu(II), and Pb(II), respectively [119]. Peanut hull, which is an abundant agricultural by-product, has also been shown to remove Cu(II) with a maximum adsorption capacity of 21.3 mg/g [120]. Wastes from other nuts have also been shown to remove heavy metals from different water sources. Several studies have investigated the ability of cashew nut shells to remove heavy metals from aqueous solutions. When evaluating the removal of Cu(II), researchers achieved significant removal (>85%) and a maximum adsorption capacity of 20 mg/g with cashew nut shells [121]. Another study evaluated the removal of Ni(II) using cashew nut shells and achieved 60–75% and a maximum adsorption capacity of 18.9 mg/g [122]. The removal of these heavy metals using cashew nut shells has been attributed primarily to its high surface area, which allows for significant number of active sites for adsorption to occur [121, 122]. Sunflower-derived adsorbents were efficiently applied against heavy in water [123]. The activated carbons generated, from chickpea (Cicer arietinum) husks by chemical treatment with KOH and K_2CO_3, efficiently removed heavy metals from aqueous solutions [37]. Pistachio hull waste also demonstrated significant removal (>98%) of Cr(VI) from various water sources, achieving a maximum adsorption capacity of 116.3 mg/g [124]. The high adsorption capacity of Cr(VI) by pistachio hull waste was attributed to the electrostatic attraction, as well as binding to various functional groups on the surface of the adsorbent [124]. Another study investigated the use of pecan shells to remove Cu(II), Pb(II), and Zn(II) by utilizing a variety of modification techniques to enhance removal, including acid, steam, and carbon dioxide activation [101]. In this study, Pb(II) was removed at the highest rate, followed by Cu(II) and Zn(II), for each type of modified pecan shell, with maximum adsorption observed for acid-activated pecan shells [101]. Tangerine peel can be used as a potential adsorbent of heavy metal ions, such as Cd, Co, Cr, Cu, Mn, Ni, Pb, and Zn, from aqueous solution [125]. Almond shells also demonstrated approximately 20–40% removal of Cr(VI) when adjusting the pH and the adsorbent dose in the solution [126]. Hazelnut shells also demonstrated effective removal of Cu(II), achieving a maximum adsorption capacity of 58.3 mg/g [127]. Groundnut shells were also used as an adsorbent in the removal of heavy metals [128]. Shukla and Pai achieved maximum adsorption capacities of 4.9, 8.05, and 11.0 mg/g for Cu(II), Ni(II), and Zn(II), respectively, with groundnut shells. These adsorption capacities were also enhanced by 40–70% with chemical modifications to the groundnut shells using reactive dye. Various fruit wastes have been shown to effectively remove heavy metals from aqueous solutions. For instance, lemon peel was shown to effectively remove Zn(II), Pb(II), Cd(II), Cu(II), and Ni(II), achieving maximum adsorption capacities of 27.9, 37.9, 54.6, 71.0, and 80.0 mg/g, respectively [129]. Orange peel also demonstrated effective heavy metal removal in a variety of studies. Ajmal et al. [99] achieved significant removal of Ni(II) (97.5%) with orange peel, along with lower removal efficiencies of Cu(II), Pb(II), Zn(II), and Cr(VI). Thirumavalavan et al. [46] investigated the adsorption of Cd(II), Cu(II), Ni(II), Pb(II), and Zn(II) with orange peel and demonstrated significant removal, achieving maximum adsorption capacities of 41.8, 63.3, 81.3, 27.1, and 24.1 mg/g, respectively. The biochars derived from agricultural wastes were utilized to remove Cd(II) and Cu(II) from aqueous [130]. Lucerne biochar had the highest Langmuir sorption capacity of Cd(II) (6.28 mg/g), and vetch-derived biochar had the highest Cu(II) sorption capacity (18.0 mg/g) at pH 5.5. Another study demonstrated similar removal of Pb(II) using orange peel, achieving a maximum adsorption capacity of 27.9 mg/g [131]. Annadurai et al. [132] also achieved much lower removal of five different heavy metals using orange peel with maximum adsorption capacities ranging from 1.9 to 7.8 mg/g in the following order of adsorption: Pb(II) > Ni(II) > Zn(II) > Cu(II) > Co(II). Significant removal of Cd(II), Cu(II), Pb(II), and Ni(II) was also achieved with chemically modified orange peel with maximum adsorption capacities of 293, 289, 476, and 162 mg/g, respectively [133, 134]. Banana peel also exhibited varying degrees of heavy metal removal in aqueous solution. Thirumavalavan et al. [129] demonstrated significant removal of a variety of heavy metals, achieving maximum adsorption capacities of 21.9, 25.9, 34.1, 52.4, and 54.4 mg/g for Zn(II), Pb(II), Cd(II), Cu(II), and Ni(II), respectively. A study conducted by DeMessie et al. [135] achieved a maximum adsorption capacity of 7.4 mg/g for Cu(II) using banana peel, which increased to 38.3 and 38.4 mg/g after the banana peel was pyrolyzed at 500 and 600°C, respectively. Banana peel, watermelon peel, and grape waste reported to be the most efficient adsorbents for the removal of heavy metal from wastewater over pH 2.0 and 5.5 [136]. Melia et al. [137] had investigated over agricultural wastes and by-products (AWBs) from grape, wheat, barley, and flax production, to reduce the concentration of Cd in contaminated water. Another study observed relatively low removal for several heavy metals using banana peel, achieving maximum adsorption capacities ranging from 2.6 to 7.9 mg/gin the following order of adsorption: Pb(II) > N i(II) > Zn(II) > Cu(II) > Co(II) [132]. Grapefruit peel was also found to be an effective adsorbent for the removal of Cd(II) and Ni(II) from aqueous solution, achieving maximum adsorption capacities of 42.1 and 46.1 mg/g, respectively [138]. The adsorption onto the grapefruit peel was attributed to the ion-exchange mechanism and, to a lesser extent, complexation with –OH functional groups [138]. Grape stalk wastes have also demonstrated the ability to remove heavy metals, achieving maximum adsorption capacities of 10.1 and 10.6 mg/g for Cu(II) and Ni(II), respectively [139]. Other types of vegetable wastes have been shown to remove heavy metals from water source. Mushroom residues were shown to be effective in the removal of heavy metals. Based on an evaluation of four different types of mushroom residues, removal efficiencies for Cu(II), Zn(II), and Hg(II) ranged from 39.7 to 81.7% [140]. Another study investigated the removal of Cd(II) and Pb(II) using three different mushrooms and achieved maximum adsorption capacities of 35.0 and 33.8 mg/g, respectively [141]. Corncob was also shown to remove heavy metals from aqueous solutions. When investigating its removal of Cd(II), researchers achieved a maximum adsorption capacity of 5.1 mg/g, along with an 4–10-fold increase in removal when the corncob was chemically modified using nitric and citric acid [142]. Moreover, corncob successfully removed Pb(II), with a maximum adsorption capacity of 16.2 mg/g. The adsorption capacity for the removal of Pb(II) using corncob increased significantly (43.4 mg/g) when the corncob was treated with sodium hydroxide. As summarized in **Table 2**, the different bioadsorbents made from different kinds of agro wastes such as orange peel, coconut shell, potato peel, rice waste, spirogyra, peanut shell, cashew nut shell, which are potentially used for the removal of various heavy metals including Cr(VI), Ni(II), Cu(II), Pb(II), Zn(II), etc. Nevertheless, the adsorption capacities

Adsorbent source	Adsorbate	Optimum pH	Removal capacity (max)	References
Orange peel	Ni(II);Cu(II);Pb(II);Zn(II);Cr(IV)	6.0	96%	Ajmal et al. [99]
Coconut shell charcoal	Cr(VI)	6.0	15.47 mg/g	Babel and Kurniawan [100]
Pecan shells	Cu(II); Pb(II); Zn(II)	4.8	~88%; ~90%; ~27%	Bansode et al. [101]
Potato peels charcoal	Cu(II)	6.0	99.8%	Amana et al. [102]
Rice husk	Cr(VI)	2.0	88.88%	Bishnoi et al. [103]
Rice hull	Cr(VI); Cu(II)	2.0; 5.5	0.17 mg/g 0.02 mg/g	Tang et al. [104]
Spirogyra	Cu(II)	5.0	133 mg/g	Gupta et al. [105]
Ecklonia maxima	Cu(II); Pb(II); Cd(II);	5.0 5.0 5.0	85–94 mg/g: 227–243 mg/g: 83.5 mg/g	Feng et al. [106]
Ulva lactuca	Cr(VI)	6.0	92%	El-Sikaily et al. [107]
Oedogonium sp.; Nostoc sp.	Pb(II)	5.0	145.0 mg/g; 93.5 mg/g	Gupta et al. [108]
Fucus serratus	Cu(II)	5.5	3.15 mmol/g	Ahmady-Asbchin et al. [109]
Rice husk	Ni(II); Zn(II); Cd(II); Mn(II); Co(II); Cu(II); Hg(II); Pb(II);	6.0	0.094 mmol/g; 0.124 mmol/g; 0.149 mmol/g; 0.151 mmol/g; 0.162 mmol/g; 0.172 mmol/g; 0.18 mmol/g; 0.28 mmol/g;	Krishnani et al. [111]

Adsorbent source	Adsorbate	Optimum pH	Removal capacity (max)	References
Coconut shell; Neem leaves; Hyacinth roots; Rice straw; Rice bran; Rice husk	Cu(II)	6.0	19.888 mg/g; 17.488 mg/g; 21.79 mg/g; 18.351 mg/g; 20.977 mg/g; 17.869 mg/g	Singha et al. [112]
Rice husk	Cr(VI)	2.0	99.5%	Georgiev et al. [113]
Rice husk	Cr(VI)	2.0	76.5%	Bansal et al. [114]
Rice husk	Cd(II)	12.0	99%	Ajmal et al. [115]
Peanut shell	Pb(II)	6.0	32.87 mg/g	Tasar et al. [116]
Peanut shell	Cr(VI)	2.0	4.48 mg/g	Ahmad et al. [117]
Peanut shell	Cu(II)	5.0	25.39 mg/g	Witek-Krowiak et al. [118]
Peanut husk	Pb(II); Cr(III); Cu(II)	4.0	4.66 mg/g; 3.02 mg/g; 3.80 mg/g	Li et al. [119]
Peanut hull	Cu(II)	5.5	21.3 mg/g	Zhu et al. [120]
Cashew nut shell	Cu(II)	5.0	20.0 mg/g	Kumar et al. [121]

Table 2.
Removal of heavy metal using agro waste adsorbent.

of adsorbent materials are dependent of the adsorption dosage, contact time, concentration, and pH. The maximum adsorption capacities were predominant at lower pH and proportional to the adsorption dosage.

4. Conclusions

Agro wastes or residues such as sugars, cellulose, minerals, and proteins are well off with nutrient composition and valuable bioactive compounds. Consequently, agro wastes having heterogeneity composition can be considered as "precursor" for other industrial processes instead of "wastes" keeping in mind sustainable development. Solid-state fermentation is a familiar approach for the production of microbial metabolites over agro waste with a low moisture content, with the advantages of a high yield concentration but only a proportionate minimum energy being needed. Various microbes have prospective to utilize the agro waste as raw materials for their growth through fermentation processes going for generation of biofuel as an alternative to faster depleting fossil fuel. The agro waste direct or active carbon generated from it can be suitable use for wastewater treatment. Vermicompost can be produced on the degradation of various agro wastes using numerous species of worms within 3–4 month time periods, which have advantages as (a) it can proceed as biofertilizers, reinstate soil nutrients, stabilizes the soil, and augmented the fertility of soil over an extended period; (b) it sorts out the social demands and recycles the waste; and (c) it is observed to be a beneficial endeavor as a circular economy. Vermicompost is found to be better option as compared with the normal composting, commonly adopted in Asian countries, owing to its enhanced nutrient contents, i.e., nitrogen, phosphorus, and potassium content. The vermicomposting also has capability to enhance the soil structure and to improve its water-holding capacity. The vermicomposting is considered to be ideal organic manure for better growth and yield of agricultural product. The agro waste can be suitably used as lowcost adsorbent for wastewater treatment. One of the most ways to generate revenue from agro waste by converting into nanocellulose and activated carbon, which have a multidisciplinary applications per today's market demands. The agro waste can have different environmental approach that leads for waste to revenue generation.

Acknowledgements

We express thanks to NIST (Autonomous) and the institute management for providing us the required infrastructure, the support, and encouragement to write and publish the book chapter.

Conflict of interest

The authors declare no conflict of interest in publication of this book chapter.

Notes/thanks/other declarations

Thanks to the authors of all references cited in this chapter whose research findings helped many ways to inculcate for this chapter.

Author details

Manabendra Patra and Duryodhan Sahu*
Department of Chemistry, National Institute of Science and Technology, Berhampur, Odisha, India

*Address all correspondence to: duryosahu@gmail.com

References

[1] Sivakumar D, Srikanth P, Ramteke PW, Nouri' J. Agricultural waste management generated by agro-based industries using biotechnology tools. Global Journal of Environmental Science and Management. 2022;**8**:281-296

[2] Martin JGP, Porto E, Correa CB, Alencar SM, Gloria EM, Cabra ISR, et al. Antimicrobial potential and chemical composition of agro industrial wastes. Journal of Natural Products. 2012;**5**:27-36

[3] Nigam PS, Gupta N, Anthwal A. Pre-treatment of agroindustrial residues. In: Nigam PS, Pandey A, editors. Biotechnology for Agro-Industrial Residues Utilization. Heidelberg: Springer; 2009. pp. 13-33

[4] El-Tayeb TS, Abdelhafez AA, Ali SH, Ramadan EM. Effect of acid hydrolysis and fungal biotreatment on agroindustrial wastes for obtainment of free sugars for bioethanol production. Brazilian Journal of Microbiology. 2012;**43**:1523-1535

[5] Motte JC, Trably E, Escudié R, Hamelin J, Steyer JP, Bernet N, et al. Total solids content: A key parameter of metabolic pathways in dry anaerobic digestion. Biotechnology for Biofuels. 2013;**6**:1-9

[6] Ferrentino G, Morozova K, Mosibo OK, Ramezani M, Scampicchio M. Biorecovery of antioxidants from apple pomace by supercritical fluid extraction. Journal of Cleaner Production. 2018;**186**:253-261

[7] Garcia-Mendoza MP, Paula JT, Paviani LC, Cabral FA, Martinez Correa HA. Extracts from mango peel byproduct obtained by supercritical CO_2 and pressurized solvent processes. LWT—Food Science and Technology. 2015;**62**:131-137

[8] Ozturk B, Parkinson C, Gonzales-Miquel M. Extraction of polyphenolic antioxidants from orange peel waste using deep eutectic solvents. Separation and Purification Technology. 2018;**206**:1-13

[9] Kelly NP, Kelly AL, Mahony JA. Strategies for enrichment and purification of polyphenols from fruit-based materials. Trends in Food Science and Technology. 2019;**83**:248-258

[10] Duque-Acevedo M, Belmonte-Urena LJ, Cortés-García FJ, Camacho-Ferre F. Agricultural waste: Review of the evolution, approaches and perspectives on alternative uses. Global Ecology and Conservation. 2020;**22**:e00902

[11] Zhang L, Zeng G, Don H, Chen Y, Zhang J, Yan M, et al. The impact of silver nanoparticles on the co-composting of sewage sludge and agricultural waste: Evolutions of organic matter and nitrogen. Bioresource Technology. 2017;**230**:132-139

[12] Qian X, Shen G, Wang Z, Guo C, Liu Y, Lei Z, et al. Co-composting of livestock manure with rice straw: Characterization and establishment of maturity evaluation system. Waste Management. 2014;**34**(2):530-535

[13] Singh S, Nain L. Microorganisms in the conversion of agricultural wastes to compost. Proceedings of the Indian National Science Academy. 2014;**80**:473-481

[14] Gusmawartati H, Yusuf M. Effect various combination of organic waste on compost quality. Journal of Tropical Soils. 2015;**20**:59-65

[15] Mastouri F, Hassandokht MR, Padasht Dehkaei MN. The effect of application of agricultural waste compost on growing media and greenhouse lettuce yield. Acta Horticulturae. 2005;**697**:153

[16] Pergola M, Persiani A, Pastore V, Palese AM, D'Adamo C, De Falco E, et al. Sustainability assessment of the green compost production chain from agricultural waste: a case study in southern Italy. Agronomy. 2020;**10**:230

[17] Aslam Z, Ahmad A, Ibrahim M, Iqbal N, Idrees M, Ali A, et al. Microbial enrichment of vermicompost through earthworm *Eisenia fetida* (Savigny, 1926) for agricultural waste management and development of useful organic fertilizer. Pakistan Journal of Agricultural Science;**58**(2021):851-861

[18] Yu Y, Li S, Qiu J, Li J, Luo Y, Guo J. Combination of agricultural waste compost and biofertilizer improves yield and enhances the sustainability of a pepper field. Journal of Plant Nutrition and Soil Science. 2019;**182**(2019):1-10

[19] Trillas MI, Casanova E, Cotxarrera L, Ordovás J, Borrero C, Avilés ML. Composts from agricultural waste and the *Trichoderma asperellum* strain T-34 suppress rhizoctonia solani in cucumber seedlings. Biological Control. 2006;**39**:32-38

[20] Karak T, Bhattacharyya P, Paul RK, Das T, Saha SK. Evaluation of composts from agricultural wastes with fish pond sediment as bulking agent to improve compost quality. Clean—Soil, Air, Water. 2013;**41**:711-723

[21] Gea FJ, Navarro MJ, Tello JC. Potential application of compost teas of agricultural wastes in the control of the mushroom pathogen verticillium fungicola. Journal of Plant Diseases and Protection. 2009;**116**:271-273

[22] Ukanwa KS, Patchigolla K, Sakrabani R, Anthony E, Mandavgane S. A review of chemicals to produce activated carbon from agricultural waste biomass. Sustainability. 2019;**11**:6204

[23] Ioannidou OA, Zabaniotou AA, Stavropoulos GG, Islam MA, Albanis TA. Preparation of activated carbons from agricultural residues for pesticide adsorption. Chemosphere. 2010;**80**:1328-1336

[24] Malik R, Ramteke DS, Wate SR. Adsorption of malachite green on groundnut shell waste based powdered activated carbon. Waste Management. 2007;**27**(9):1129-1138

[25] Kadirvelu K, Namasivayam C. Activated carbon from coconut coir pith as metal adsorbent: Adsorption of Cd(II) from aqueous solution. Advances in Environmental Research. 2003;**7**:471-478

[26] Prahas D, Kartika Y, Indraswati N, Ismadji S. Activated carbon from jackfruit peel waste by H_3PO_4 chemical activation: Pore structure and surface chemistry characterization. Chemical Engineering Journal. 2008;**140**:32-42

[27] Ioannidou O, Zabaniotou A. Agricultural residues as precursors for activated carbon production. Renewable and Sustainable Energy Reviews. 2007;**11**:1966-2005

[28] Buhani S, Luziana F, Rilyanti M, Sumadi S. Production of adsorbent from activated carbon of palm oil shells coated by Fe_3O_4 particle to remove crystal violet in water. Desalination and Water Treatment. 2019;**171**:281-293

[29] Razi MAM, Al-Gheethi A, Al-Qaini M, Yousef A. Efficiency of activated carbon from palm kernel shell for treatment of grey-water. Arab

Journal of Basic and Applied Sciences. 2018;**25**(3):103-110

[30] Keppetipola NM, Dissanayake M, Dissanayake P, Karunarathne B, Dourges MA, Talaga D, et al. Graphite-type activated carbon from coconut shell: A natural source for eco-friendly non-volatile storage devices. RSC Advances. 2021;**11**:2854

[31] Omokafe SM, Adeniyi AA, Igbafen EO, Oke SR, Olubambi PA. Fabrication of activated carbon from coconut shells and its electrochemical properties for supercapacitors. International Journal of Electrochemical Science. 2020;**15**:10854-10865

[32] Feng P, Li J, Wang H, Zhiqiang X. Biomass-based activated carbon and activators: Preparation of activated aarbon from corncob by chemical activation with biomass pyrolysis liquids. ACS Omega. 2020;**5**:24064-24072

[33] Saleem J, Bin Shahid U, Hijab M, Mackey H, McKay G. Production and applications of activated carbons asadsorbents from olive stones. Biomass Conversion and Biorefinery. 2019;**9**:775-802

[34] Lia Z, Hanafy H, Zhang L, Sellaoui L, Netto MS, Oliveira MLS, et al. Adsorption of congo red and methylene blue dyes on an ashitaba waste and a walnut shell-based activated carbon from aqueous solutions: Experiments, characterization and physical interpretations. Chemical Engineering Journal. 2020;**388**:124263

[35] Ash B, Satapathy D, Mukherjee PS, Nanda B, Gumaste JL, Mishra BK. Characterization and application of activated carbon prepared from coir pith. Journal of Scientific and Industrial Research. 2006;**65**:1008-1012

[36] Putra Negara DNK, Nindhia TGT, Septiadi WN. Surface properties and adsorption capacities of rice bran-activated carbon. Journal of Mechanical Engineering Science and Technology. 2020;**4**(2):115-124

[37] Ozsin G, Kilic M, Apaydın-Varol E, Putun AE. Chemically activated carbon production from agricultural waste of chickpea and its application for heavy metal adsorption: equilibrium, kinetic, and thermodynamic studies. Applied Water Science. 2019;**9**:56

[38] Chuayjumnong S, Karrila S, Jumrat S, Pianroj Y. Activated carbon and palm oil fuel ash as microwave absorbers for microwave-assisted pyrolysis of oil palm shell waste. RSC Advances. 2020;**10**:32058

[39] Tay T, Ucar S, Karagöz S. Preparation and characterization of activated carbon from waste biomass. Journal of Hazardous Materials. 2009;**165**:481-485

[40] Nunes AA, Franca AS, Oliveira LS. Activated carbons from waste biomass: An alternative use for biodiesel production solid residues. Bioresource Technology. 2009;**100**:1786-1792

[41] Chang KL, Lin JH, Chen ST. Adsorption studies on the removal of pesticides (Carbofuran) using activated carbon from rice straw agriculture waste. World Academy of Science, Engineering and Technology. 2011;**76**:348

[42] Chang KL, Hsieh JF, Ou BM, Chang MH, Hsieh WY, Lin JH, et al. Adsorption studies on the removal of an endocrine-disrupting comound (bisphenol A) using activated carbon from rice straw agriculture waste. Separation Science and Technology. 2012;**47**:1514

[43] Isoda N, Rodrigues R, Silva A, Gonçalves M, Mandelli D, Figueiredo FCA, et al. Optimization

of preparation conditions of activated carbon from agriculture waste utilizing factorial design. Powder Technology. 2014;**256**:175-181

[44] Köseoğlu E, Akmil-Basar C. Preparation, structural evaluation and adsorptive properties of activated carbon from agricultural waste biomass. Advanced Powder Technology. 2015;**26**:811-818

[45] Mahamad MN, Zaini MAA, Zakaria ZA. Preparation and characterization of activated carbon from pineapple waste biomass for dye removal. International Biodeterioration & Biodegradation. 2015;**102**:274

[46] Baysal M, Bilge K, Yılmaz B, Papila M, Yürüm Y. Preparation of high surface area activated carbon from waste-biomass of sunflower piths: Kinetics and equilibrium studies on the dye removal. The Journal of Environmental Chemical Engineering. 2018;**6**:1702

[47] Patil A, Suryawanshi UP, Harale NS, Patil SK, Vadiyar MM, Luwang MN, et al. Adsorption of toxic Pb(II) on activated carbon derived from agriculture waste (mahogany fruit shell): Isotherm, kinetic and thermodynamic study. International Journal of Environmental Analytical Chemistry. 2020. DOI: 10.1080/03067319.2020.1849648

[48] Xue H, Wang X, Xu Q, Dhaouadi F, Sellaoui L, Seliem MK, et al. Adsorption of methylene blue from aqueous solution on activated carbons and composite prepared from an agricultural waste biomass: A comparative study by experimental and advanced modeling analysis. Chemical Engineering Journal. 2022;**430**:132801

[49] Mateo S, Peinado S, Morillas-Gutiérrez F, La Rubia MD, Moya AJ. Nanocellulose from agricultural wastes: Products and applications—A review. Processes. 2021;**9**:1594

[50] Fillat Ú, Wicklein B, Martín-Sampedro R, Ibarra D, Ruiz-Hitzky E, Valencia C, et al. Assessing cellulose nanofiber production from olive tree pruning residue. Carbohydrate Polymers. 2018;**179**:252-261

[51] Rambabu N, Panthapulakkal S, Sain M, Dalai A. Production of nanocellulose fibers from pinecone biomass: Evaluation and optimization of chemical and mechanical treatment conditions on mechanical properties of nanocellulose films. Industrial Crops and Products. 2016;**83**:746-754

[52] Dos Santos RM, Neto WPF, Silvério HA, Martins DF, Dantas NO, Pasquini D. Cellulose nanocrystals from pineapple leaf, a new approach for the reuse of this agro-waste. Industrial Crops and Products. 2013;**50**:707-714

[53] Johar N, Ahmad I, Dufresne A. Extraction, preparation and characterization of cellulose fibres and nanocrystals from rice husk. Industrial Crops and Products. 2012;**37**:93-99

[54] Siqueira G, Tapin-Lingua S, Bras J, da Silva Perez D, Dufresne A. Morphological investigation of nanoparticles obtained from combined mechanical shearing and enzymatic and acid hydrolysis of sisal fibers. Cellulose. 2010;**17**:1147-1158

[55] Ohwoavworhua F, Adelakun T. Non-wood fibre production of microcrystalline cellulose from Sorghum caudatum: Characterisation and tableting properties. Indian Journal of Pharmaceutical Sciences. 2010;7:295

[56] Fortunati E, Luzi F, Jiménez A, Gopakumar D, Puglia D, Thomas S, et al. Revalorization of sunflower stalks

as novel sources of cellulose nanofibrils and nanocrystals and their effect on wheat gluten bionanocomposite properties. Carbohydrate Polymers. 2016;**149**:357-368

[57] Ferreira FV, Mariano M, Rabelo SC, Gouveia RF, Lona LMF. Isolation and surface modification of cellulose nanocrystals from sugarcane bagasse waste: From a micro- to a nano-scale view. Applied Surface Science. 2018;**436**:1113-1122

[58] Lu P, Hsieh YL. Preparation and characterization of cellulose nanocrystals from rice straw. Carbohydrate Polymers. 2012;**87**:564-573

[59] do Nascimento DM, Almeida JS, Vale MDS, Leitão RC, Muniz CR, de Figueirêdo MCB, et al. A comprehensive approach for obtaining cellulose nanocrystal from coconut fiber. Part I: Proposition of technological pathway. Industrial Crops and Products. 2016;**93**:66-75

[60] de Carvalho Mendes CA, Ferreira NMS, Furtado CRG, de Sousa AMF. Isolation and characterization of nanocrystalline cellulose from corn husk. Materials Letters. 2015;**148**:26-29

[61] Zheng D, Zhang Y, Guo Y, Yue J. Isolation and characterization of nanocellulose with a novel shape from walnut (*Juglans regia* L.) shell agricultural waste. Polymers. 2019;**11**:1130

[62] Sijabat EK, Nuruddin A, Aditiawati P, Purwasasmita BS. Optimization on the synthesis of bacterial nano cellulose (BNC) from banana peel waste for water filter membrane applications. Materials Research Express. 2020;**7**:055010

[63] Panpatte DG, Jhala YK. Agricultural waste: A suitable source for biofuel production. In: Rastegari AA et al., editors. Prospects of Renewable Bioprocessing in Future Energy Systems, Biofuel and Biorefinery Technologies. Vol. 10. Springer Nature Switzerland AG; 2019

[64] Santa-Maria M, Ruiz-Colorado AA, Cruz G, Jeoh T. Assessing the feasibility of biofuel production from lignocellulosic banana waste in rural agricultural communities in Peru and Colombia. Bioenergy Research. 2013;**6**:1000-1011

[65] Srivastava AK, Agrawal P, Rahiman A. Delignification of rice husk andproduction of bioethanol. International Journal of Innovative Research in Science, Engineering and Technology. 2014;**2**:10187-10194

[66] Singh A, Bajar S, Narsi R, Bishnoi NR. Enzymatic hydrolysis of microwave alkali pretreated rice husk for ethanol production by *Saccharomyces cerevisiae*, *Scheffersomyces stipitis* and their co-culture. Fuel. 2014;**116**:699-702

[67] Chukwuma SE, Olawale O, Ikechukwu NE, Chigozie ME, Ositadinma CU, Ephraim NA. Enhanced availability of biofuel and biomasscomponents in aspergillus Niger and aspergillus fumigatus treated rice husk. European Scientific Journal. 2014;**10**:1857-7881

[68] Naqvi SR, Ali I, Nasir S, Ali S, Taqvi A, Atabani AE, et al. Assessment of agro-industrial residues for bioenergy potential by investigating thermo-kinetic behavior in a slow pyrolysis process. Fuel. 2020;**278**:118259

[69] Buenrostro-Figueroa J, Tafolla-Arellano JC, Flores-Gallegos AC, Rodríguez-Herrera R, De la Garza-Toledo H, Aguilar CN. Native yeasts for alternative utilization of

overripe mango pulp for ethanol production. Revista Argentina de Microbiología. 2018;**50**:173-177

[70] Mihajlovski K, Buntić A, Milić M, Rajilić-Stojanović M, Dimitrijević-Branković S. From agricultural waste to biofuel: Enzymatic potential of a bacterial isolate streptomyces fulvissimus cks7 for bioethanol production. Waste and Biomass Valorization. 2021;**12**:165-174

[71] Najafi G, Ghobadian B, Tavakoli T, Yusaf T. Potential of bioethanol production from agricultural wastes in Iran. Renewable and Sustainable Energy Reviews. 2009;**13**:1418-1427

[72] Taghizadeh-Alisaraei A, Assar HA, Ghobadian B, Motevali A. Potential of biofuel production from pistachio waste in Iran. Renewable and Sustainable Energy Reviews. 2017;**72**:510-522

[73] Yuliansyah AT, Hirajim T, Kumagai S, Sasaki K. Production of solid biofuel from agricultural wastes of the palm oil industry by hydrothermal treatment. Waste and Biomass Valorization. 2010;**1**:395-405

[74] Laobussararak B, Chulalak-sananukul W, Chavalparit O. Comparison of bacterial and yeast ethanol fermentation Yield from Rice straw. Advanced Materials Research. 2012;**347-353**:2541-2544

[75] Kumar AK, Parikh BS, Shah E, Liu LZ, Cotta MA. Cellulosic ethanol production from green solvent-pretreated rice straw. Biocatalysis and Agricultural Biotechnology. 2016;**7**:14-23

[76] Sasaki K, Matsuda F, Hasunuma T, Ogino C, Urairi M, Yoshida K, et al. Ability of a perfluoropolymer membrane to tolerate by-products of ethanol fermentation broth from dilute acid-pretreated rice straw. Biochemical Engineering Journal. 2013;**70**:135-139

[77] Jain RK, Ghosh D, Agrawal D, Suman SK, Pandey D, Vadde VT, et al. Ethanol production from rice straw using thermotolerant Kluyveromyces sp. IIPE453. Biomass Conversion and Biorefinery. 2015;**5**:331-337

[78] de Assis Castro RC, Roberto IC. Effect of nutrient supplementation on ethanol production in different strategies of saccharification and fermentation from acid pretreated rice straw. Biomass and Bioenergy. 2015;**78**:156-163

[79] Mahajan C, Chadha BS, Nain L, Kaur A. Evaluation of glycosyl hydrolases from thermophilic fungi for their potential in bioconversion of alkali and biologically treated *Parthenium hysterophorus* weed and rice straw into ethanol. Bioresource Technology. 2014;**163**:300-307

[80] Sasaki K, Tsuge Y, Sasaki D, Hasunuma T, Sakamoto T, Sakihama Y, et al. Optimized membrane process to increase hemicellulosic ethanol production from pretreated rice straw by recombinant xylose-fermenting Saccharomyces cerevisiae. Bioresource Technology. 2014;**169**:380-386

[81] Momayez F, Karimi K, Horváth IS. Enhancing ethanol and methane production from rice straw by pretreatment with liquid waste from biogas plant. Energy Conversion and Management. 2018;**178**:290-298

[82] Molaverdia M, Karimia K, Mirmohamadsadeghia S, Galbec M. High titer ethanol production from rice straw via solid-state simultaneous saccharification and fermentation by Mucor indicus at low enzyme loading.

Energy Conversion and Management. 2019;**182**:520-529

[83] Lü J-L, Zhou P-J. Ethanol production from microwave-assisted $FeCl_3$ pretreated rice straw. Energy Sources, Part A: Recovery, Utilization, and Environmental Effects. 2015;**37**(21):2367-2374

[84] Okamoto K, Nitta Y, Maekawa N, Yanase H. Direct ethanol production from starch, wheat bran and rice straw by the white rot fungus *Trametes hirsute*. Enzyme and Microbial Technology. 2011;**48**:273-277

[85] Karimi K, Emtiazi G, Taherzadeh MJ. Ethanol production from dilute acid pretreated rice straw by simultaneous saccharification and fermentation with *Mucor indicus*, *Rhizopus oryzae*, and *Saccharomyces cerevisiae*. Enzyme and Microbial Technology. 2006;**40**:138e144

[86] Kaur J, Chugh P, Soni R, Soni SK. A low-cost approach for the generation of enhanced sugars and ethanol from rice straw using in-house produced cellulase-hemicellulase consortium from a. niger P-19. Bioresource Technology Reports. 2020;**11**:100469

[87] Sharma M, Khan AA, Puri SK, Tuli DK. Wood ash as a potential heterogeneous catalyst for biodiesel synthesis. Biomass and Bioenergy. 2012;**41**:94-106

[88] Uprety BK, Chaiwong W, Ewelike C, Rakshit SK. Biodiesel production using heterogeneous catalysts including wood ash and the importance of enhancing byproduct glycerol purity. Energy Conversion and Management. 2016;**115**:191-199

[89] Betiku JE, Etim AO, Pereao O, Ojumu TV. Two-step conversion of neem (*Azadirachta indica*) seed oil into fatty methyl esters using a heterogeneous biomass-based catalyst: An example of cocoa pod husk. Energy & Fuels. 2017;**31**:6182-6193

[90] Vadery V, Narayanan BN, Ramakrishnan RM, Cherikkallinmel SK, Sugunan S, Narayanan DP, et al. Room temperature production of jatropha biodiesel over coconut husk ash. Energy. 2014;**70**:588-594

[91] Sarma K, Kumar P, Aslam M, Chouhan APS. Preparation and characterization of *Musa balbisiana* Colla underground stem nano-material for biodiesel production under elevated conditions. Catalysis Letters. 2014;**144**:1344-1353

[92] Aslam M, Saxena P, Sarma AK. Green technology for biodiesel production from *Mesua ferrea* L. seed oil. Energy & Environmental Science. 2014;**4**(2):11-21

[93] Gohain M, Devi A, Deka D. *Musa balbisiana* Colla peel as highly effective renewable heterogeneous base catalyst for biodiesel production. Industrial Crops and Products. 2017;**109**:8-18

[94] Betiku E, Ajala SO. Modeling and optimization of *Thevetia peruviana* (yellow oleander) oil biodiesel synthesis via *Musa paradisiacal* (plantain) peels as heterogeneous base catalyst: A case of artificial neural network vs. response surface methodology. Industrial Crops and Products. 2014;**53**:314-322

[95] Etim AO, Betiku E, Ajala SO, Olaniyi PJ, Ojumu TV. Potential of ripe plantain fruit peels as an ecofriendly catalyst for biodiesel synthesis: Optimization by artificial neural network integrated with genetic algorithm. Sustainability. 2018;**10**:707

[96] Betiku E, Akintunde AM, Ojumu TV. Banana peels as a biobase catalyst for fatty acid methyl esters production using napoleon's plume (*Bauhinia monandra*) seed oil: A process parameters optimization study. Energy. 2016;**103**:797-806

[97] Odude VO, Adesina AJ, Oyetunde OO, Adeyemi OO, Ishola NB, Etim AO, et al. Application of agricultural waste-based catalysts to transesterification of esterified palm kernel oil into biodiesel: A case of banana fruit peel versus cocoa pod husk. Waste and Biomass Valorization. 2019;**10**:877-888

[98] Igwe JC, Ogunewe DN, Abia AA; Competitive adsorption of Zn(II), Cd(II) and Pb(II) ions from aqueous and non-aqueous solution by maize cob and husk African Biotechnol J; **4**(2005):1113-1116.

[99] Ajmal M, Rao R, Ahmad R, Ahmad J. Adsorption studies on *Citrus reticulata* (fruit peel of orange) removal and recovery of Ni(II) from electroplating waste water. Journal of Hazardous Materials. 2000;**79**:117-131

[100] Babel S, Kurniawan TA. Cr(VI) removal from synthetic wastewater using coconut shell charcoal and commercial activated carbon modified with oxidizing agents and/or chitosan. Chemosphere. 2004;**54**(7):951-967

[101] Bansode PR, Losso JN, Marshall WE, Rao RM, Portier RJ. Adsorption of metal ions by pecan shell-based granular activated carbons. Bioresource Technology. 2003;**89**:115-119

[102] Amana T, Kazi AA, Sabri MU, Banoa Q. Potato peels as solid waste for the removal of heavy metal copper(II) from waste water/industrial effluent. Colloids and Surfaces B: Biointerfaces. 2008;**63**:116-121

[103] Bishnoi NR, Bajaj M, Sharma N, Gupta A. Adsorption of Cr(VI) on activated rice husk carbon and activated alumina. Bioresource Technology. 2003;**91**(3):305-307

[104] Tang P, Lee CK, Low KS, Zainal Z. Sorption of Cr(VI) and Cu(II) in aqueous solution by ethylenediamine modified rice hull. Environmental Technology. 2003;**24**:1243-1251

[105] Gupta VK, Rastogi A, Saini VK, Jain N. Biosorption of copper(II) from aqueous solutions by Spirogyra species. Journal of Colloid and Interface Science. 2006;**296**:59-63

[106] Fenga D, Aldrich C. Adsorption of heavy metals by biomaterials derived from the marine alga *Ecklonia maxima*. Hydrometallurgy. 2004;**73**:1-10

[107] El-Sikaily A, El Nemr A, Khaled A, Abdelwehab O. Removal of toxic chromium from wastewater using green alga ulva lactuca and its activated carbon. Journal of Hazardous Materials. 2007;**148**:216-228

[108] Gupta VK, Rastogi A. Biosorption of lead(II) from aqueous solutions by non-living algal biomass Oedogonium sp. and Nostoc sp.—A comparative study. Colloids and Surfaces B: Biointerfaces. 2008;**64**:170-178

[109] Ahmady-Asbchin S, Andre's Y, Ge'rente C, Le Cloirec P. Biosorption of Cu(II) from aqueous solution by *Fucus serratus*: Surface characterization and sorption mechanisms. Bioresource Technology. 2008;**99**:6150-6155

[110] Deliyanni EA, Peleka EN, Matis KA. Removal of zinc ion from water by sorption onto iron-based nanoadsorbent.

Journal of Hazardous Materials. 2007;**141**:176-184

[111] Krishnani KK, Meng X, Christodoulatos C, Boddu VM. Biosorption mechanism nine different heavy metals onto biomatrix from rice husk. The Journal of Hazardous Materials. 2008;**153**:1222-1234

[112] Singha B, Das SK. Adsorptive removal of Cu(II) from aqueous solution and industrial effluent using natural/agricultural wastes. Colloids and Surfaces. B, Biointerfaces. 2013;**107**:97-106

[113] Georgieva VG, Tavlieva MP, Genieva SD, Vlaev LT. Adsorption kinetics of Cr(VI) ions from aqueous solutions onto black rice husk ash. Journal of Molecular Liquids. 2015;**208**:219-226

[114] Bansal M, Garg U, Singh D, Garg VK. Removal of Cr(VI) from aqueous solutions using pre-consumer processing agricultural waste: A case study of rice husk. Journal of Hazardous Materials. 2009;**162**:312-320

[115] Ajmal M, Rao RA, Anwar S, Ahmad J, Ahmad R. Adsorption studies on rice husk: Removal and recovery of Cd(II) from wastewater. Bioresource Technology. 2003;**86**:147-149

[116] Taşar S, Kaya F, Özer A. Biosorption of lead(II) ions from aqueous solution by peanut shells: Equilibrium, thermodynamic and kinetic studies. Journal of Environmental Chemical Engineering. 2014;**2**:1018-1026

[117] Ahmad A, Ghazi ZA, Saeed M, Ilyas M, Ahmad R, Khattaka AM, et al. A comparative study of the removal of Cr(VI) from synthetic solution using natural biosorbents. New Journal of Chemistry. 2017;**41**:10799-10807

[118] Witek-Krowiak A, Szafran RG, Modelski S. Biosorption of heavy metals from aqueous solutions onto peanut shell as a low-cost biosorbent. Desalination. 2011;**265**:126-134

[119] Li Q, Zhai J, Zhang W, Wang M, Zhou J. Kinetic studies of adsorption of Pb(II), Cr(III) and Cu(II) from aqueous solution by sawdust and modified peanut husk. Journal of Hazardous Materials. 2007;**141**:163-167

[120] Zhu CS, Wang LP, Chen WB. Removal of Cu(II) from aqueous solution by agricultural by-product: Peanut hull. Journal of Hazardous Materials. 2009;**168**:739-746

[121] Kumar PS, Ramalingam S, Sathyaselvabala V, Kirupha SD, Sivanesan S. Removal of copper(II) ions from aqueous solution by adsorption using cashew nut shell. Desalination. 2011;**266**:63-71

[122] Kumar PS, Ramalingam S, Kirupha SD, Murugesa A, Vidhyadevi T, Sivanesan S. Adsorption behavior of nickel(II) onto cashew nut shell: Equilibrium, thermodynamics, kinetics, mechanism and process design. Chemical Engineering Journal. 2011;**167**:122-131

[123] Anastopoulos I, Ighalo JO, Igwegbe CA, Giannakoudakis DA, Triantafyllidis KS, Pashalidis I, et al. Sunflower-biomass derived adsorbents for toxic/heavy metals removal from wastewater. Journal of Molecular Liquids. 2021;**342**:117540

[124] Moussavi G, Barikbin B. Biosorption of chromium(VI) from industrial wastewater onto pistachio hull waste biomass. Chemical Engineering Journal. 2010;**162**:893-900

[125] Abdić Š, Memić M, Šabanović E, Sulejmanović J, Begić S. Adsorptive

removal of eight heavy metals from aqueous solution by unmodified and modified agricultural waste: Tangerine peel. International journal of Environmental Science and Technology. 2018;**15**:2511-2518

[126] Dakiky M, Khamis M, Manassra A, Mer'eb M. Selective adsorption of chromium(VI) in industrial wastewater using low-cost abundantly available adsorbents. Advances in Environmental Research. 2002;**6**:533-540

[127] Demirbas E, Dizge N, Sulak MT, Kobya M. Adsorption kinetics and equilibrium of copper from aqueous solutions using hazelnut shell activated carbon. Chemical Engineering Journal. 2009;**148**:480-487

[128] Shukla SR, Pai RS. Adsorption of Cu(II), Ni(II) and Zn(II) on dye loaded groundnut shells and sawdust. Separation and Purification Technology. 2005;**43**:1-8

[129] Thirumavalavan M, Lai Y, Lin L, Lee J. Cellulose-based native and surface modified fruit peels for the adsorption of heavy metal ions from aqueous solution. Journal of Chemical & Engineering Data. 2010;**55**:1186-1192

[130] Bandara T, Xu J, Potter ID, Frankse A, Chathurika JBAJ, Tang C. Mechanisms for the removal of Cd(II) and Cu(II) from aqueous solution and mine water by biochars derived from agricultural wastes. Chemosphere. 2020;**254**:126745

[131] Abdelhafez AA, Li J. Removal of Pb(II) from aqueous solution by using biochars derived from sugar cane bagasse and orange peel. Journal of the Taiwan Institute of Chemical Engineers. 2016;**61**:367-375

[132] Annadurai G, Juang RS, Lee DJ. Adsorption of heavy metals from water using banana and orange peels. Water Science and Technology. 2002;**47**:185-190

[133] Feng N, Guo X, Liang S. Adsorption study of copper(II) by chemically modified orange peel. Journal of Hazardous Materials. 2009;**164**:1286-1292

[134] Feng N, Guo X, Liang S, Zhu Y, Liu J. Biosorption of heavy metals from aqueous solutions by chemically modified orange peel. Journal of Hazardous Materials. 2011;**185**:49-54

[135] De Messie B, Sahle-Demessie E, Sorial G. Cleaning water contaminated with heavy metal ions using pyrolyzed biochar adsorbents. Separation Science and Technology. 2015;**50**:2448-2457

[136] Lorenzo M, Antonella G, Astolfi ML, Congedo R, Masotti A, Canepari S. Efficiency evaluation of food waste materials for the removal of metals and metalloids from complex multi-element solutions. Materials. 2018;**11**:334

[137] Melia M, Busquets PR, Ray S, Cundy AB. Agricultural wastes from wheat, barley, flax and grape for the efficient removal of Cd from contaminated wate. RSC Advances. 2018;**8**:40378-40386

[138] Asadollahzadeh M, Torab-Mostaedi M, Hemmati A, Khosravi A. Equilibrium, kinetic, and thermodynamic studies for biosorption of cadmium and nickel on grapefruit peel. Journal of the Taiwan Institute of Chemical Engineers. 2013;**44**:295-302

[139] Villaescusa I, Fiol N, Martinez M, Miralles N, Poch J, Serarols J. Removal of copper and nickel ions from aqueous solutions by grape stalks wastes. Water Research. 2004;**38**:992-1002

[140] Li X, Zhang D, Sheng F, Qing H. Adsorption characteristics of copper(II), zinc(II) and mercury(II) by four kinds of immobilized fungi residues. Ecotoxicology and Environmental Safety. 2018;**147**:357-366

[141] Vimala R, Das N. Biosorption of cadmium(II) and lead(II) from aqueous solutions using mushrooms: A comparative study. Journal of Hazardous Materials. 2009;**168**:376-382

[142] Tan G, Yuan H, Liu Y, Xiao D. Removal of lead from aqueous solution with native and chemically modified corncob. Journal of Hazardous Materials. 2010;**174**:740-745

Chapter 4

Perspective Chapter: Industrial Waste Landfills

Olawale Theophilus Ogunwumi and Lukumon Salami

Abstract

Wastes are generated as a result of anthropogenic activities. The rapid industrialization of human society in the twenty-first century has led to an increase in the generation of industrial wastes that have negatively impacted humans and the environment. While industrial operations and techniques have improved globally, leading to a higher standard of living, economic prosperity, and healthcare delivery, industries have continued to produce waste on a scale never before seen. This chapter discussed industrial wastes, waste generation, and industries involved, waste disposal, landfilling as a disposal method, effects of waste disposal, modern techniques in industrial waste management, landfill sustainability, and regulations.

Keywords: industrial waste, waste generation, waste disposal, effect of disposal, waste management, landfill sustainability and regulations

1. Introduction

Industrialization, which meets the needs of a rapidly increasing global population, is the backbone of economic development and human welfare in any society. However, the proliferation of industries and industrial activities in many countries, especially developed countries in Europe and America, has led to environmental pollution and eco-deterioration around the developed world. Industries in less developed countries of Africa and the Sahel are no different as the impacts of their activities on the environment continue to produce far more deterioration due to poor pollution management and control. For instance, Teku [1] and Firdissa et al. [2] stated that in Ethiopia, rapid industrialization has led to the generation of industrial wastes including hazardous wastes, and improper management of the vast amount of these wastes is one of the most critical environmental problems in Addis Ababa. One of the main concerns of environmental engineers and activists in Nigeria, especially in the state of Lagos with a small land mass in comparison with its large population, is the ever-expanding solid waste landfills and their attendant environmental degradation caused by leachate formation (**Figure 1**), which can contaminate arable agricultural land, surface water, and aquifers [3].

Figure 1.
Leachate flowing across the road at a solid waste dumpsite in Lagos, Nigeria [3].

2. Industrial wastes and their types

Industrial wastes are unwanted and residual materials produced by industrial activities and may include any product discarded as useless during such manufacturing operations as metal deformation, metal casting, sheet metal forming, polymer processing, machining, finishing, assembly, foundry, steam generation, and coal-fire electrification, construction works, textile manufacture, pulp and paper mills, and mining operations to mention a few [4]. These wastes are basically generated in large quantities when compared to municipal wastes and mostly at every stage in the manufacture of products, which turns raw materials into finished goods that are sold or distributed [4]. Manufacturing processes commonly generate all forms of waste including gaseous, liquid, and solid wastes [4].

Industrial waste products have particularly dangerous properties such as toxicity, ignitability, corrosivity, or reactivity [5], can negatively impact human health and the environment [6] or pollute nearby soil, adjacent to water bodies or contaminate groundwater, lakes, streams, rivers, and coastal waters [3]. At a typical landfill site, industrial waste is often mixed with municipal waste, which makes accurate assessments of industrial waste produced difficult [2]. In the United States, for instance, an estimation gave more than 7.5 billion tons of industrial waste produced annually, as of 2017 [2]. Problems associated with generation of industrial waste have forced most countries and municipalities to enact legislation to deal with the situation [2]. While strictness and compliance with industrial waste pollution legislation may vary from place to place, enforcement of such legislation is often an issue [2]. The different types of industrial waste generated according to the industrial sector are shown in **Table 1** below.

2.1 Classes of industrial wastes

Industries generate various kinds of waste depending on their manufacturing processes and the finished product being produced, which are classified according

S/N	Industrial sector	Description/industrial processes	Typical type of waste
1.	Mining and quarrying	Extraction, beneficiation, and processing of minerals	Tailings, phosphogypsum, muds, solid rock, and slag
2.	Energy	Electricity, gas, steam, and air conditioning supply	Boiler slag, fly ash, bottom ash, sludge, particulates, and used oils
3.	Manufacturing	Chemical Food Textile Paper	Sludge, spent catalyst, chemical solvents, reactive waste, alkali, used oils, ash, and particulate waste Packaging, carton, and plastic Pigments, peroxide, textile wastes, alkali, organic stabilizer, chemical solvents, heavy metals, and sludge Chemical solvents, sludge, wood waste, and alkali
4.	Construction	Demolition and construction activities	Glass, plaster, cinder blocks, concrete, mansory, gypsum, wood shingles, asphalt, metals, and slate
5.	Wastewater services	Water supply, collection, and treatment	Spent sludge and adsorbents

Table 1.
Type of industrial waste according to industry sector [7].

to their state/nature, degradation potential and toxicology, and hydrocarbon content [8]. Generally speaking, industrial waste can be classified as biodegradable and non-biodegradable wastes. The different classes and more specifically, types of industrial wastes generated during any manufacturing process are described in the following section.

2.1.1 Biodegradable and nonbiodegradable industrial wastes

Industries generate wastes that may or may not be decomposed by microorganisms (such as bacteria, fungi, algae, protozoa) when deposited in landfills.

2.1.1.1 Biodegradable industrial wastes

Biodegradable industrial wastes are produced from industrial processes, which generate decomposable material in which decomposition is caused by microbial activities and the material is then converted to gas and water. These kinds of industrial wastes are similar to municipal wastes and are usually generated by food processing and agro-allied industries, slaughterhouses and dairy industries, etc. [8]. These wastes are nonhazardous, do not require special kind of treatment, and are mostly solid. They include such wastes as animal bones, fur, wheat, animal skin, leather, wool, discarded fruits, and so on [8].

2.1.1.2 Nonbiodegradable industrial wastes

These are industrial wastes that cannot be decomposed into gases and water. They are generated by industries such as fertilizer, chemical and petroleum, drugs and pharmaceutical, mechanical, dye industries, nuclear power plants, polymer processing, construction, foundry, metal, and steel plants [8]. Examples of this waste are polyethylene terephthalate (PET) plastics, fly ash, synthetic fibers, glass, gypsum, and radioactive wastes [8]. Most landfills consist of nonbiodegradable industrial waste due to their nondecomposing nature [8]. Other types of industrial wastes can further be divided into the following types based on their nature.

2.2 Types of industrial wastes based on their nature

2.2.1 Industrial solid waste

Industrial solid waste (**Figure 2**) is produced by specific manufacturing processes and can either be biodegradable or be nonbiodegradable but mostly, it is neither municipal nor hazardous wastes [8]. It encompasses a wide range of materials of varying environmental toxicity [8]. Typically, this range of waste materials would include paper, packaging materials, waste from food processing and agrochemical, oils, solvents, resins, paints and sludges, glass, ceramics, stones, metals, plastics, rubber, leather, wood, cloth, straw, abrasives, etc. [8]. Generally, according to Ref. [5] there are about 43 different industrial solid waste material ranges that can be categorized into 5 groups, which include the following:

- Solid chemicals.
- Drugs and food.
- Textile and apparel including clothes, leather, and laundry.
- Energy such as coal processing, power plants including thermal and nuclear.
- Pulp and paper, steel, rubber.

The major generators of industrial solid wastes are the thermal power plants producing coal ash, integrated iron and steel mills producing blast furnace slag and steel melting slag, nonferrous industries like aluminum, zinc, and copper producing red mud and tailings, sugar industries generating press mud, pulp, and paper industries producing lime and fertilizer and allied industries producing gypsum, and food and pharmaceutical industries producing solid food and biomedical wastes, respectively [9]. It is important to note that there is a distinction between solid and liquid industrial wastes because they are handled very differently and are generated on largely different scales [9].

2.2.2 Industrial hazardous waste

The term industrial hazardous waste refers to any semisolid, solid, sludge, or liquid waste generated from such industry activities as mining, medical services, and public services including waste incineration plants, waste recycling plants, which contains chemicals, pathogens, radiation, heavy metals, or other toxins thereby posing a significant threat to human health and the environment [10]. Hazardous waste generated in industries may be listed or characteristic (**Figure 3**). While listed industrial hazardous wastes are produced by specific industries or generated from common manufacturing processes and discarded as waste commercial products, characteristic

Figure 2.
A stack of waste paper produced in the pulp and paper industry [7].

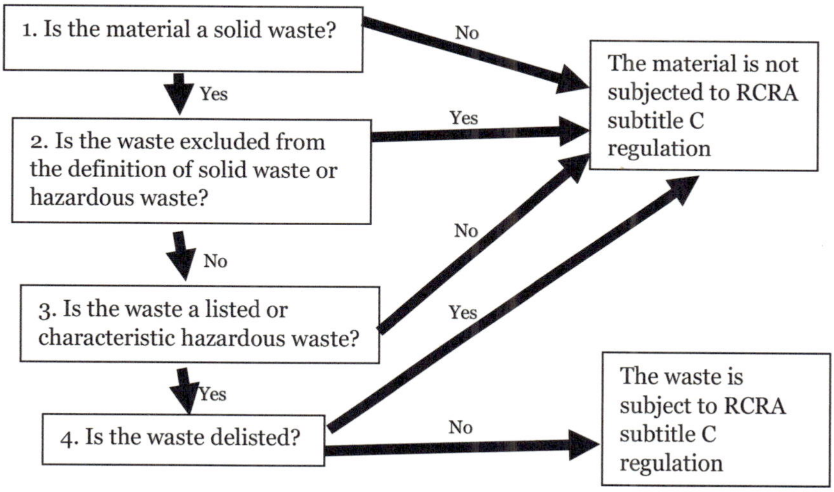

Figure 3.
Identification of industrial hazardous wastes [11].

industrial hazardous wastes show one or more characteristic properties such as ignitability, corrosivity, reactivity, or toxicity; some hazardous wastes are referred to as mixed wastes when they contain both hazardous and radioactive components [11]. All forms of hazardous wastes including listed, characteristic, and mixed are regulated under both Resource Conservation and Recovery Act (RCRA) of 1976 and Atomic Energy Act (AEA) of 1946 [12].

Common examples of hazardous wastes are chlorinated aliphatic hydrocarbons, multisource leachate, petroleum refinery sludge, explosive solvent, inorganic pigment, organic and inorganic chemicals (listed wastes), flammable fluids, aqueous acids and bases, explosive and toxic gases (characteristic wastes), low-level mixed waste (LLMW), high-level mixed waste (HLW), and transuranic waste (MTRU) (mixed wastes) [11].

Nonhazardous industrial wastes may be generated from industrial activities associated with the production of iron and steel, pulp and paper, glass, electric power generation, electronic and electrical products, construction and concrete, and they do not meet Environmental Protection Agency's (EPA) definition of hazardous waste [12]. Although they are not considered to pose immediate risks to human health, they can still be harmful to the environment as a result of methane emission during decomposition [12]. Some common industrial nonhazardous waste is ash, sludge, antifreeze, grinding dust, etc. Electronic wastes (or E-wastes) are generally nonhazardous; however, there are a few that may be considered hazardous. Electronic devices and components that can be disposed of as nonhazardous include zinc plating, aluminum, copper, gold, etc. [12].

2.2.3 Industrial chemical waste

These are solid, liquid, or gaseous materials that are hazardous in nature and are in excess, unused, or unwanted during the manufacture of a commercial product. Chemical wastes in the industry may result from expired or extraneous materials when certain industrial processes or products have become extinct [13]. Also, materials that are contaminated with flammable liquids such as acetone, acetonitrile; leachate toxic substances like heavy metals, pesticides; corrosive substances like potassium hydroxide pellets and hydrochloric acids; reactive substances like cyanides, oxidizers, sulfides, explosives, benzoyl peroxide, and toxic substances like mutagenic, carcinogenic, chloroform, ethidium bromide; other chemical wastes are pyritic slag, acidic slag, alkali slag, salt mud, mud from chemical production kettle, residues of refining or distillation; all these require careful handling and disposal [13].

2.2.4 Industrial radioactive waste

Radioactive or nuclear wastes are generated by industrial activities that use or produce radioactive materials. These wastes are hazardous as they emit radioactive particles, which are risky and can pose huge threat to the environment if not properly handled [14]. Radioactive wastes can arise from the commercial operations of nuclear reactors, hospitals, research facilities, and fuel processing plants [14]. Examples of radioactive wastes are low-level wastes, which include common everyday materials such as rags, plastic bags, paper, and packaging materials; high-level wastes such as used fuel from nuclear reactors and wastes from reprocessing of used-up nuclear fuel; transuranic wastes and uranium or thorium mill tailings [14].

3. Industrial waste generation

Industrial wastes are unavoidably produced during manufacturing process of commercial products by different industrial activities from agriculture to extraction of natural resources. The type of waste generated and its characteristic depend on the industry sector [15] and the inherent industrial process employed to produce finished products in such industry. Generally, according to Ref. [16] the world currently generates about 2.02 billion tons of solid waste annually; this figure is projected to be on steady increase (**Figure 4**) and may reach 3.40 billion tons by 2050.

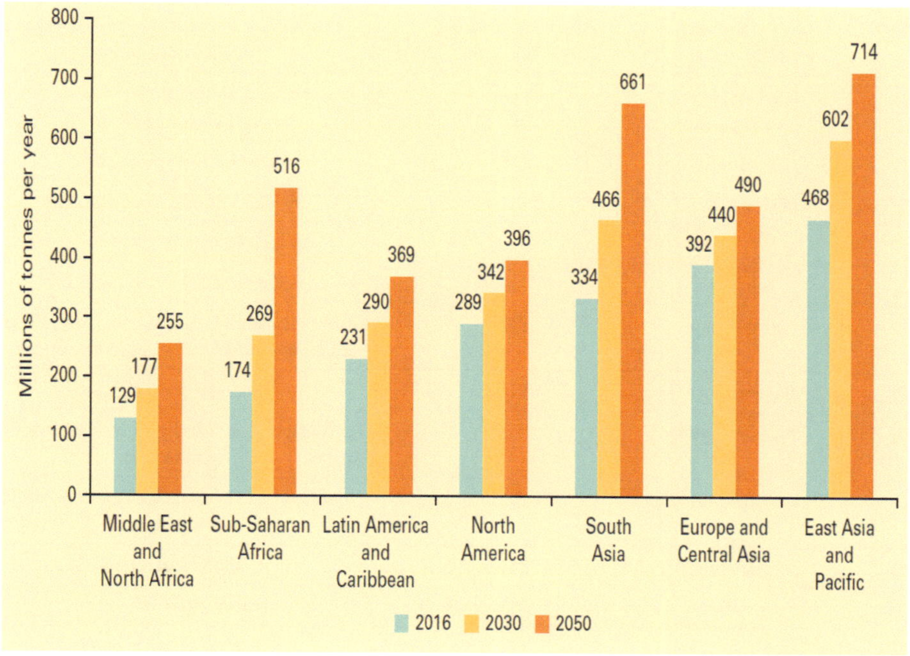

Figure 4.
Projected global waste generation by region (million tons/year) [16].

The volume of waste generated in industries around the world has equally been on the rise over a long period of time. In the United States for instance, the amount of hazardous waste produced by industries has increased from about 4.6 million tons annually after World War II, to an estimated 57 million tons by 1975 [17]. By the 1990s, the volume of waste produced by American industries had increased to approximately 260 million tons [15]. **Table 2** shows the amount of waste in million tons generated in industries in different regions of the world in 2011.

Solid wastes from industries are generated by construction activities, cement and limestone production, mining and mineral extraction, steel and iron ore

S/N	ITEMS	EUROPE	AMERICAS	AFRICA and MIDDLE EAST	ASIA PACIFIC	GLOBAL
1	Industrial wastes generated (Million tons)	1933	915	921	5357	9126
3	Countries	Germany, United Kingdom, France, Russia, and Bulgaria	United States, Brazil, Canada, Chile, and Columbia	South Africa, Saudi Arabia, United Arab Emirates, Egypt and Tunisia	China, Japan, India, South Korea, and Australia	

Table 2.
Volume of industrial wastes generated in different regions of the world in 2011 [18].

S/N	Type of industrial solid waste	Amount produced (million tons)	Product manufacturing process/waste production source
1.	Steel and blast furnace residue	35.0	Conversion of pig iron to steel and in manufacturing of iron rod
2.	Brine mud	0.02	Caustic soda production
3.	Copper slag	0.0164	By-product of copper smelting
4.	Fly ash	70.0	From coal-based thermal power plants
5.	Kiln dust	1.6	From cement production plants
6.	Lime sludge	3.0	In sugar, paper, fertilizer, soda ash, calcium carbide, production, and tannery industry
7.	Mica scrap waste	0.05	In mica mining.
8.	Phosphogypsum	4.5	In phosphoric acid and ammonium phosphate production
9.	Red mud or bauxite	3.0	In mining and extraction of aluminum from bauxite
10.	Coal washery dust	3.0	From coal mines
11.	Iron tailings	11.25	In production of iron ore
12.	Limestone waste	50.0	In limestone production/quarry

Table 3.
Industrial solid waste generated from typical industrial manufacturing processes [9].

production, and solid chemical production as can be seen in **Table 3**, which shows industrial solid wastes produced from typical manufacturing processes [9]. From a global standpoint, plastic wastes account for the most type of solid waste generated

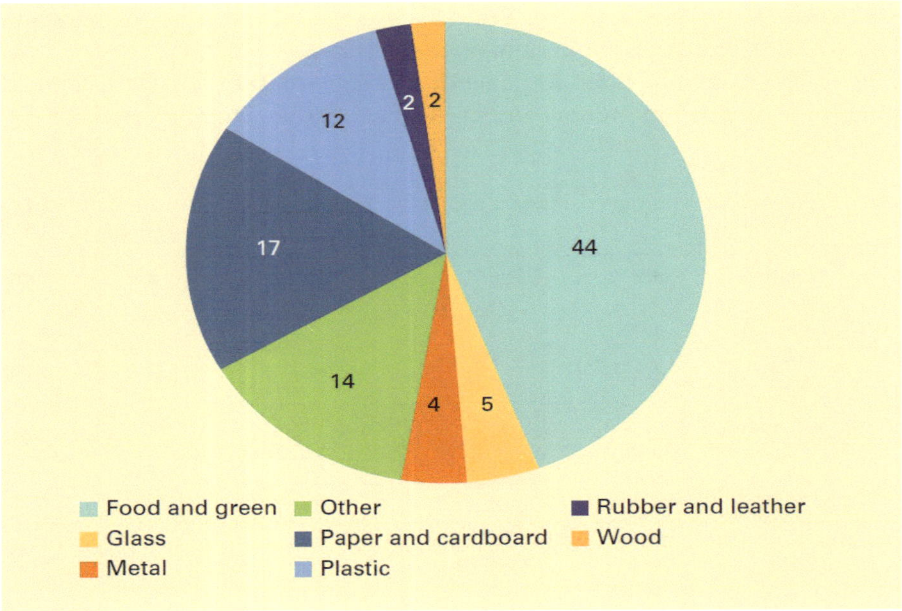

Figure 5.
Global composition of solid waste generated annually (percent) [16].

S/N	MANUFACTURED PRODUCTS	ASSOCIATED HAZARDOUS WASTE
1	Pesticides	Organic chlorine and phosphate compounds
2	Textiles	Heavy metals, dyes, organic chlorine compounds, and solvents
3	Medicines and drugs	Organic solvents and residues, heavy metals such as mercury and zinc
4	Leather	Heavy metals, organic solvents
5	Oil, Petroleum products	Oil, phenols, organic compounds, heavy metals, etc.
6	Metals	Heavy metals, fluorides, cyanides, acid and alkaline cleaners, solvents, and pigments.
7	Paints	Heavy metals, pigment, solvent, and organic residues

Table 4.
Industrial hazardous wastes associated with the production of certain commercial products [19].

(**Figure 5**) by both industry and household annually [16]. Industrial hazardous wastes, on the other hand, can be generated by industries manufacturing pharmaceuticals and drugs, paint, textile, pesticides and agrochemicals, plastic and polymer, tannery, and petroleum and petrochemical industries [17]. In **Table 4**, typical hazardous wastes generated from the manufacture of commercial, finished products are presented.

Chemical wastes from industries are generated by industrial processes, which produce chemicals such as toluene, acetone, benzene, urea, potash, sulfuric and hydrochloric acids, utilized in different sectors such as agriculture, military, service

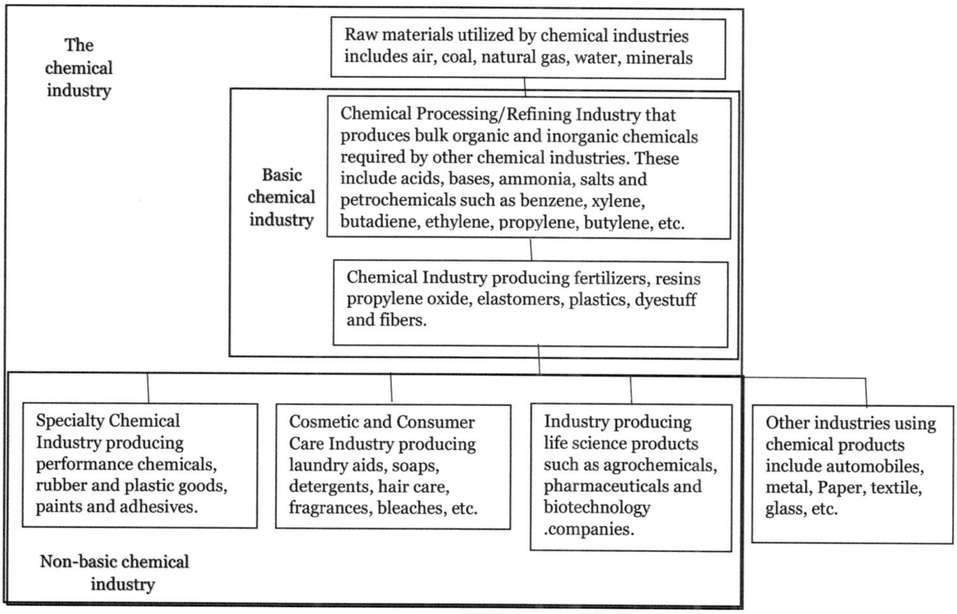

Figure 6.
The different chemical industries that produce industrial chemical waste [19].

S/N	Radioactive waste	Industrial process/source of waste	Associated radiation/half-life (years)
1.	Very low-level waste	Medical procedures	^{131}I/8.028 days
2.	Low-level waste	Industrial waste from nuclear power plants; medical and research wastes such as paper, plastic, and glass	^{3}H/2.32 ^{60}Co/5.27
3.	High-level radioactive waste	Used fuel from nuclear power reactors; liquid waste from the reprocessing of spent fuel	^{90}Sr/29.78 ^{137}Cs/30.07
4.	Mixed waste	Weapon production waste and some research wastes	^{239}Pu/24,100 ^{241}Pu/14.4
5.	Transuranic waste	Weapon production waste and mixed transuranic waste	^{238}Pu/87.7 ^{241}Am/432.7
6.	Naturally occurring radioactive material (NORM) wastes	Crude transport; scale buildup on pipe walls carrying petroleum products	^{226}Ra/1599 ^{228}Ra/5.76
7.	Uranium or thorium mill tailings waste	Milling for rare earth metal extraction	^{230}Th/75,400

Table 5.
Industrial radioactive wastes produced from various sources and processes [21].

industry, construction, consumer goods manufacture, and cosmetics [13]. Chemical wastes are also produced by paint industries and textile industries [20]. A block diagram of the general overview of the chemical industry is presented in **Figure 6** below.

Industrial radioactive waste is produced by such industries as nuclear power generation, mining, military and defense, medicine, and research institutes [14]. **Table 5** shows the various industrial radioactive wastes generated by different industrial operations.

4. Industrial waste disposal and landfilling

The millions of tons of waste generated by manufacturing industries are collected, removed, and disposed of to prevent pollution and ensure environmental protection [9]. However, disposal is only carried out after applying waste reuse and waste recycling management techniques. Other reasons for waste disposal in industries are as follows: to generate revenue from waste and to create employment opportunities [9]. Since waste has different characteristics, the method of disposal varies from one waste to another. When industrial waste is generated, pre-disposal activity like segregation is often carried out to separate waste according to type and nature. A lot of wastes are a mixture of hazardous and nonhazardous wastes and their contents may even be liquid [9]. By segregating toxic constituents in wastes, isolating liquid fractions, and keeping hazardous streams away from nonhazardous wastes, industries can save substantial amounts of money on disposal or get new opportunities in waste recycling and reuse [9]. Major waste disposal methods are described in the following section.

4.1 Industrial waste disposal methods

Industrial waste disposal can be carried out by methods such as waste recycling, waste composting, waste incineration, and landfilling.

4.1.1 Recycling of industrial waste

Waste recycling, also called beneficial waste usage, involves reusing waste materials generated from industrial processes to produce same or completely different finished products [22]. This has the advantage of reducing the need for producing too many new items and, in turn, reduces disposal at landfills which ultimately is beneficial to the environment [23]. There are a lot of opportunities for waste recycling in the industry and some wastes produced can be reused for road construction and repair, bridge construction, production of consumer goods, etc. [22]. Common recyclable industrial solid waste materials are coal combustion products, demolition and construction materials (**Figure 7**), scrap tires, and foundry sand [22]. Some of the reasons for recycling wastes by any industry are as follows: to reduce carbon footprint and conserve resources, to earn and save money, to save resources and energy when recycling machines are used, to reduce overall disposal cost, to improve company's reputation and create job opportunities and to build brand as ecofriendly, etc.

4.1.2 Industrial waste composting

Industrial waste composting or commercial composting is a large-scale recycling of organic, biodegradable waste materials in order to produce natural soil enhancers or fertilizers used in farming [24]. The composts formed in this waste disposal method (the process is shown in **Figure 8**) are then sold to agricultural centers, farms, or plant nurseries. Typically composting industrial wastes is achieved by first, collecting waste from food processing industries, grocery stores, large restaurants, and other commercial facilities and processing them at a compost processing plant (compost machine) where foreign materials are initially removed and the food wastes mixed with sludge [24].

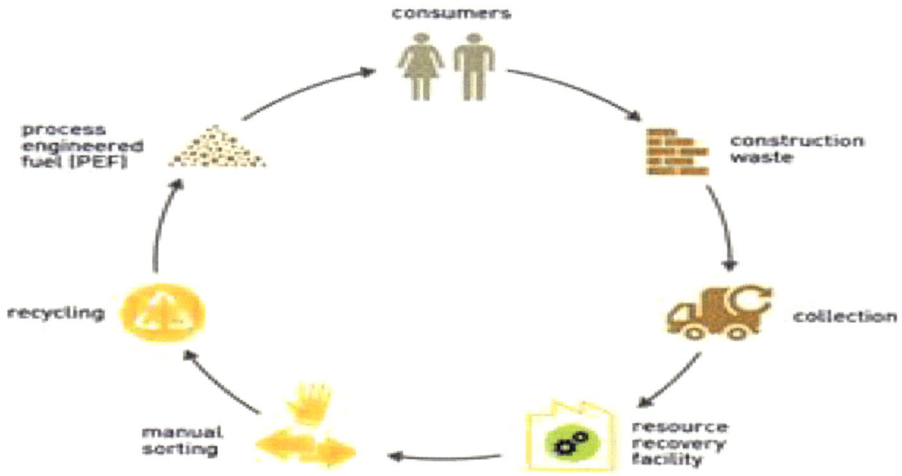

Figure 7.
Industrial construction waste recycling [23].

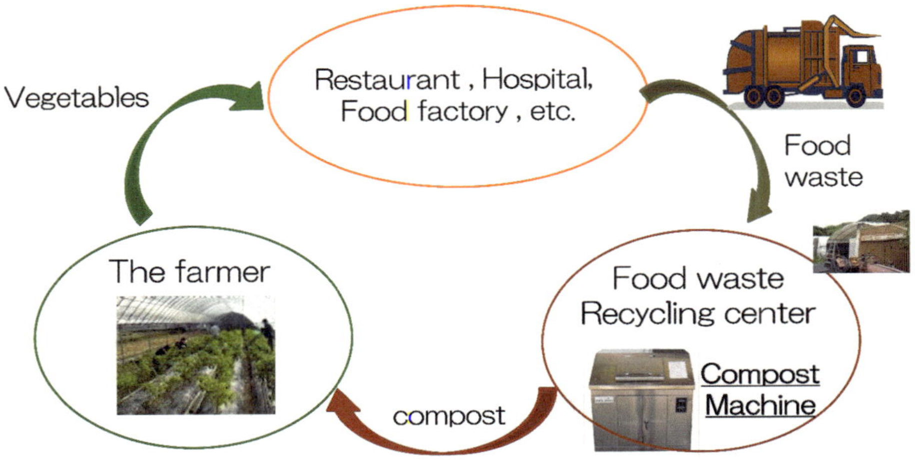

Figure 8.
Food waste recycling/compost production process [25].

4.1.3 Industrial waste incineration

Incineration of wastes from industries (**Figure 9**) is a disposal technique that involves burning combustible materials in wastes thereby converting them to flue gas (stack gas) and ash and producing heat energy that can be utilized for power generation [27]. Industrial waste incineration is a thermal waste disposal method and can be regarded as a waste-to-energy technology. The purpose of this waste disposal technique is to reduce waste volume and save landfilling costs, to prevent the release of chemical and toxic substances to the environment, and to generate revenue from energy recovered from waste combustion, which can be used for heating or electricity generation [27]. Incineration is ideal for disposing of industrial wastes from the medical, food processing, chemical, and nuclear power sectors where high temperatures are used to destroy pathogens, toxins, and hazardous materials from waste generated in these sectors [28]. There are, however, growing concerns about the environmental

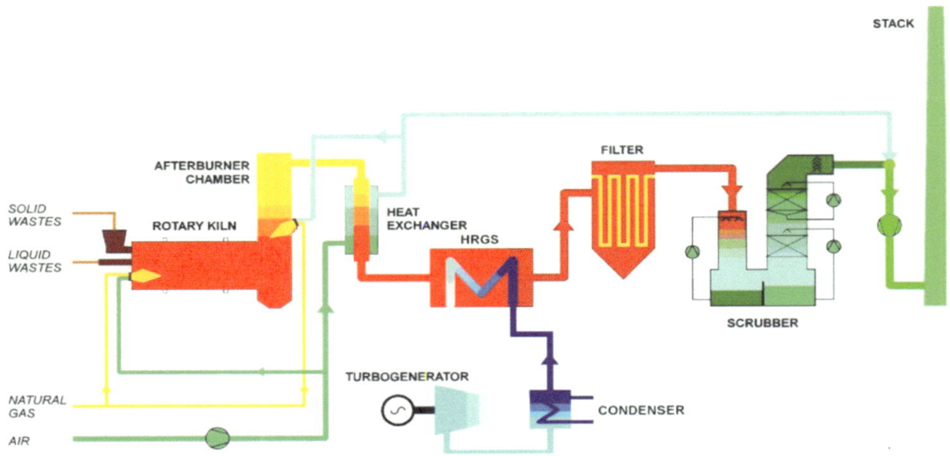

Figure 9.
Industrial waste incineration process [26].

effects of burning industrial wastes in incinerators due to the release of toxic gases during combustion.

4.1.4 Landfilling and industrial waste landfills

Landfilling is the most common, globally practiced solid waste disposal technique that involves storing large metric tons of waste in an excavated space. Landfills can store municipal, industrial, green, or hazardous wastes. According to Ref. [29] in a 2010 study, landfilling is responsible for over 30% of waste disposed of or managed in the European Union. The researchers noted that all the countries of the European Union generated about 550 kg of municipal solid waste per household in 2010 and over 400 kg of such waste was managed per person out of which almost 40% was landfilled and lesser percentage incinerated, recycled, or composted. Landfilling is practiced due to its advantages that include large volume storage, inexpensive operational procedures, and maintenance cost when compared to a disposal method such as incineration [29] but it has also been observed to pose huge environmental risks and endangerment as a result of the formation of biogas produced by the fermentation of organic matter in decomposing solid waste and leachate formed by the percolation of water through waste [3].

It is important to distinguish between an engineering landfill, whose main purpose is to ensure safety by reducing harm from accumulated wastes and allowing safe decomposition [30] and an open dumpsite (**Figure 10**). Landfills are setup and controlled by the municipal, state, or federal governments and as such are built at designated places while dumpsites on the other hand are dug by individuals, households, or communities without special consideration for setup site [30]. Landfills are covered with compact soil on regular basis to prevent offensive odors (due to biogas formation) from polluting the surrounding area; dumpsites are usually not covered with soil and so, cause air pollution [30]. Also, monitoring is an important part of any

Figure 10.
A typical open dumpsite in Lagos, Nigeria [3].

landfill operation as the drainage system and the liners are monitored by engineers to ensure no seepage of polluted liquid (leachate) formed within the landfill, and enters underground water; landfills are also designed with gas collection system and treatment plants for the liquid and gas produced; dumpsites, however, do not have liners or require monitoring and they are not setup with the treatment plants [30].

There are four different types of landfills that can be used for waste disposal. They include municipal solid waste, construction and demolition debris (C&DD) waste, hazardous waste (secure landfill), and green waste landfills [31]. The most common of the four types is the municipal solid waste landfill. This landfill is used to dispose of wastes from households and residential areas such as used tissues and waste cardboard. They have strict regulations with regard to disposal and operations, groundwater monitoring, landfill lining, closing practices, etc. [31]. The C&DD landfill is an industrial waste landfill used for disposing of majorly construction or demolition solid waste debris such as concrete and bricks. They may also be used to dispose other types of industrial solid waste such as asphalt, gypsum, metals, lumbers. C&DD landfills are used as material recovery facilities where wastes are sorted according to their usefulness and reusable wastes are separated from nonreusable ones [31]. Hazardous waste landfills are the most structured and regulated of all the landfill types, and this is due to the characteristic of waste disposed here. They may also be regarded as industrial waste landfill as they can be used to dispose industrial hazardous wastes such as pelletized potassium hydroxide, cyanides, oxidizers, sulfides, ethidium bromide. They store hazardous wastes in such a way as to prevent accidental discharge into the environment [31]. Lastly, green waste landfills are storage areas for organic waste materials to decompose naturally into compost. They are used to dispose biodegradable waste such as food, fruits, vegetables, garden, and other agricultural wastes [31]. A global comparison (**Figure 11**) of all the solid waste management techniques discussed above showed that the open dumpsite is readily utilized for disposal of waste. This can be attributed to the fact that dumpsites are setup any and everywhere regarded for proper site selection [16].

4.1.4.1 Industrial waste landfills

The large volume of wastes generated from manufacturing processes, especially solid wastes, are disposed at an industrial landfill. Although separate industrial landfills can be used to dispose construction, demolition wastes (C&DD landfill), and hazardous wastes, in practice, these landfills are combined in one facility as industrial waste landfill. Because of the nature of industrial wastes, landfills are designed with all the qualities and characteristics of engineering, sanitary landfill with modern design, stricter regulations, and mode of operations. Some of the design requirements of an industrial waste landfill involve provision of a compacter or compressor and plastic covering for top wastes after each operation, using an impervious double liner at landfill bottom [32], gas and leachate collection systems, groundwater quality monitoring system, etc. [33].

4.1.4.2 Industrial waste landfill design structure

One of the major activities carried out before the design of any engineering landfill is site selection. Important consideration must be given to the location of the facility to reduce operational impact on immediate surrounding and environs. Usually, tests are carried out to determine such factors as surface water vicinity, groundwater

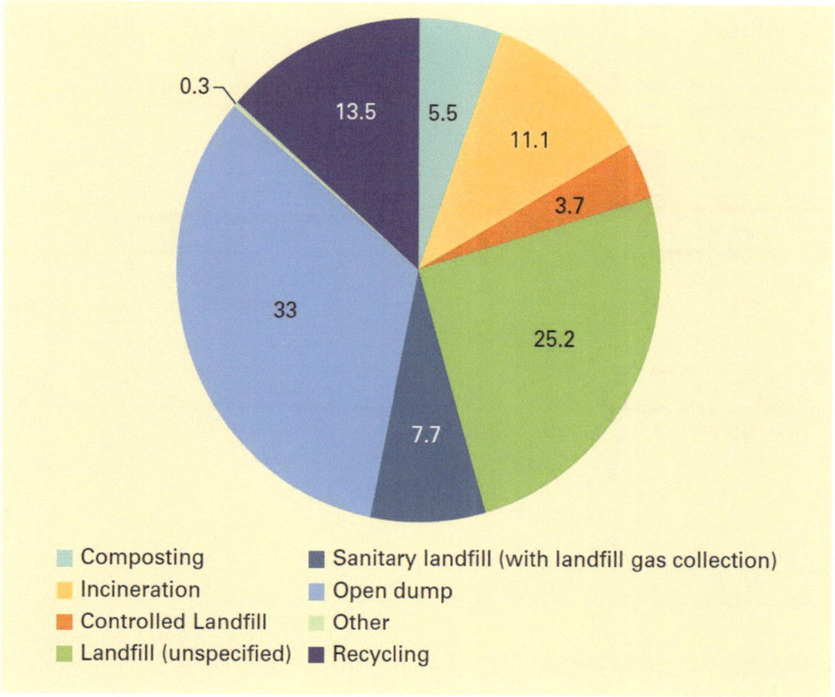

Figure 11.
Global comparison of solid waste management techniques (percent) [16].

depth, land slope, soil permeability, land elevation, soil stability, stratification and lithology faults, flood susceptibility, type of land use, urbanization and land settlement, cultural and protected site, road, airport and railway proximity, wind direction, pipe- and powerlines proximity [34]. The general layout of the landfill facility according to [9] should be made to comprise the following units as a minimum:

- Operational building with amenities.
- Operational area.
- Vehicle and instrument workshop.
- Control systems.
- Illumination, roads, fencing, trenches, etc.
- Truck/vehicle weighbridge.
- Laboratory for sample analysis.

The engineering design involved in the construction of an industrial waste landfill requires constructing layers of different sizes in a known volume of excavated space (**Figure 12**). The smallest layer is usually located at the bottom of the landfill, while the largest layer is at the top: to prevent collapse of the surrounding area and indeed, the landfill [31]. These layers facilitate decomposition of waste materials and

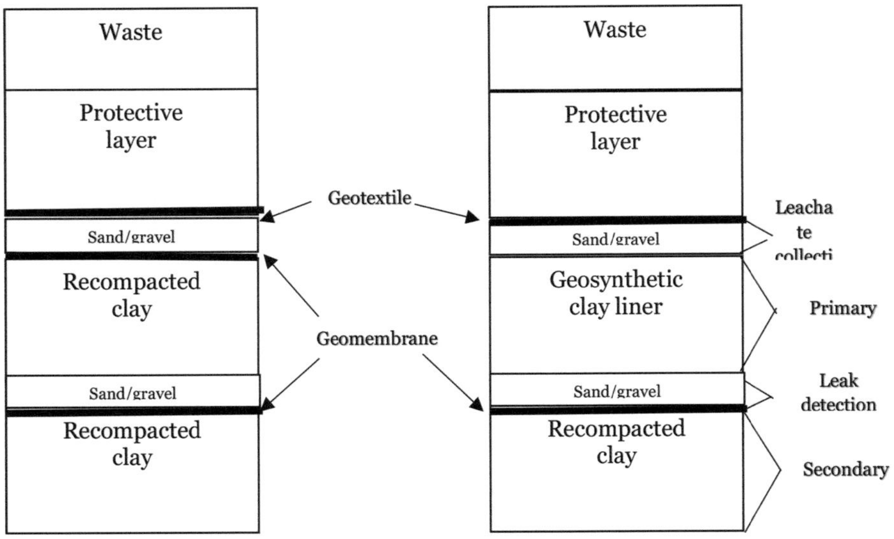

Figure 12.
Schematic representation of a typical industrial waste landfill structure [35].

entrapment of toxic gases released from within the landfill [31]. The different layers of an industrial waste landfill are described below.

4.1.4.2.1 First or bottom layer

This is the liner system. Industrial waste landfills are mostly designed with double-liner systems especially when they dispose hazardous waste (**Figure 13**), although some are built with a single-liner system (**Figure 14**); these are ideal for disposing C&DD wastes. The liner system acts as a barrier that isolates landfill contents from contact with the environment thereby preventing pollution. Common liner materials include clay, geomembranes, or flexible membrane liners (FML) made from plastic materials such as polyvinyl chloride (PVC) and high-density polyethylene (HDPE),

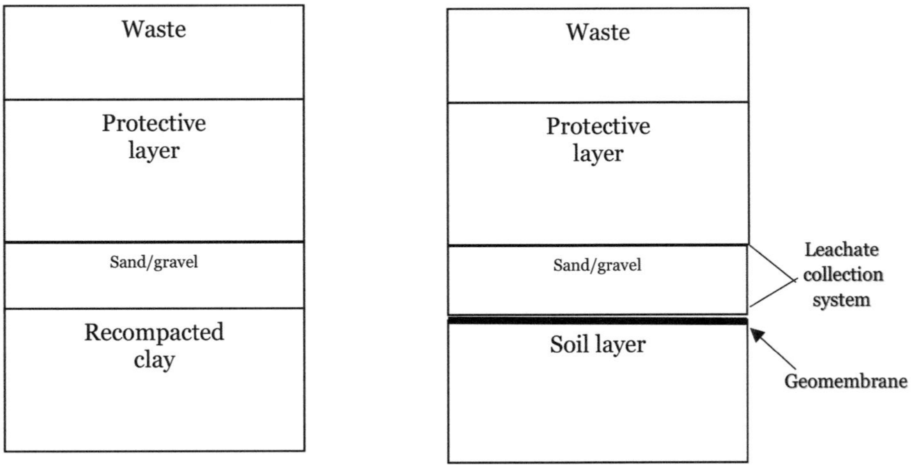

Figure 13.
Block diagram of a double-liner system mostly used in hazardous waste landfills [33].

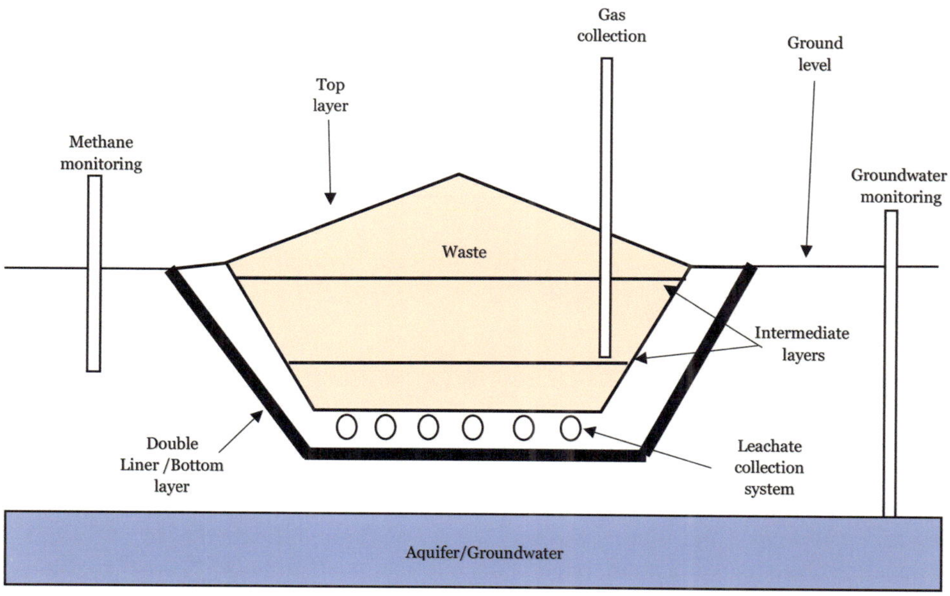

Figure 14.
Single-liner system used in C&DD and industrial solid waste landfills [33].

geotextiles, geosynthetic clay liners (GCL), and geonets [33]. The liner material is applied at the bottom of landfill to prevent seepage of liquids. Sometimes when clay is used as liner, HDPE can be applied on top of the clay as reinforcement [31].

4.1.4.2.2 Second layer

This is the drainage system. The drainage system controls toxic liquid or leachate produced from decomposing waste materials combined with rainwater runoff or snow in the landfill. The drainage layer helps in draining leachate to avoid contact with the liner system, which may be corroded by the toxicity level of leachate [31]. Leachate should never seep past the liner layer because of contamination of groundwater and soil. In order to reduce the risk of contamination, the landfill is designed with perforated pipes on top of the liners to collect all leachates at the bottom from where it is channeled to leachate treatment plant, where it is treated and then reused [31]. Stormwater and snow which may not have mixed with decomposing matter and had seeped from the top surface into the landfill are also drained away in the drainage layer.

4.1.4.2.3 Third layer

This is the gas collection system. Just as toxic liquids are produced in landfill, toxic gases (usual biogas) are also released through natural decomposition processes [31]. Methane that is a major constituent of biogas is produced in landfills. Since methane is toxic and known to contribute to global warming, it is prevented from being released into the atmosphere where it causes public health damage, by the landfill design [31]. The design of the landfill is such that gas extraction pipes are fitted into the gas collection system, which entraps methane and transports it to the gas treatment plant for electricity and power generation [31].

4.1.4.2.4 Fourth layer

This is the topmost and largest layer of the landfill that contains the disposed waste itself. A compressor is used to reduce the size of waste disposed on daily basis to avoid occupying too much space [31]. Also, compact soil is applied on top of the compressed waste to keep away windblown debris, offensive odors produced within the landfill, and pests [31].

5. Landfill sustainability

Over time, preventing, recycling, or re-using wastes have not been found to take away the need for a landfill in any waste management system as not all wastes can be recycled or recovered under all circumstances [32]. Oftentimes, the quantity of wastes to be disposed far exceeds the capacity for recovery, recycling, or incineration making the need for landfills even more inevitable, as they (landfills) have been used to dispose large quantity of wastes [32]. All of these indicate that in a typical recycling and recovery waste management practice, landfilling is a "go-to" method of waste management [32]. In landfill designs, impermeable liners are the standard for preventing environmental pollution [32]. Liners may stay intact for a long period, up to 50 or even 100 years but eventually, they will fail; this necessitates the need for landfill aftercare, which may not be sustainable [32]. Landfills should, therefore, be managed and operated so that future generations do not have to worry about their adverse effect on the environment; they should be managed in a sustainable way.

A sustainable landfill is one which attains stable conditions within a short period of time and reaches a state where the undisturbed contents no longer pose environmental risk to human health and the environment [32]. At this stage, the landfill is said to be in a state of completion and aftercare can be discontinued [32]. In sustainable landfills, waste materials are safely absorbed into the surrounding environment, whether or not they have been treated and landfill gases (LFGs) are controlled so as to minimize environmental impact [32]. In order to achieve landfill completion, waste pretreatment or bioreactor operations can be carried out in the landfill facility. Common chemicals used for pretreatment include aluminum sulfate, ferrous sulfate, ferric chloride, and ferric chlorosulfate [3]. Bioreactor operations are employed for waste treatment in a landfill [33]. A landfill bioreactor facilitates the degradation of waste by microbial activity, which may be achieved by controlling parameters such as moisture, oxygen, nutrients, pH, and temperature [33]. Since water limits microbial activity in a landfill, leachate recirculation can be used to create landfill bioreactor [33]. Recirculation increases the moisture content of the waste in the landfill thereby promoting waste degradation. Landfill sustainability can be measured (**Figure 13**) in terms of environmental, economic, and social impacts of the landfill on host community [33].

6. Environmental effect of poor disposal of industrial waste

Improper or poor waste disposal is injurious to man and the environment. Indiscriminate disposal allows for uncontrolled and unhygienic decomposition of organic, toxic, and hazardous waste materials [32]. When wastes decompose uncontrollably outside an engineering landfill facility, air pollution, microorganisms and pests infestation, and bad esthetics are the result [36]. Industrial solid wastes contain toxic and hazardous chemicals whose concentrations usually exceed permissible levels [9] and when released into the environment, these chemicals may result in physical, biological, and chemical disruptions of the ecosystem and soil fertility [36]. Plastic wastes, for instance, have been linked to endocrine disruptions in humans and animals due to the slow release of bisphenol A at disposal [37]. Toxic chemicals and other harmful substances in leachate from dumpsites or poorly managed landfills may seep into the soil and pollute groundwater [3]. Also, some industrial hazardous and radioactive wastes mixed with municipal solid wastes (cardboard and scraps) at dumpsites produce noxious gases and dioxins when burned [36]. These gases are carcinogenic and have the potential to cause disease and eventually, death from exposure [36].

Effects of improper disposal of industrial wastes on humans and animals cannot be felt without a pathway (route) for contact or exposure. Some of the exposure routes include: (i) exposure by skin contact, which is the direct contact of humans with dioxins that can bring about irritation [9]. Pollutants like corrosive acids can destroy the skin by a single, one-time exposure, while others such as organic solvents may cause damage by repeated exposure [9]; (ii) exposure by inhalation is the easiest source of occupational workplace contact with pollutants and the most difficult to control [9]. Air pollutants can affect the respiratory tract, causing damage to the lungs, bronchi/bronchioles, larynx, and trachea [9]; (iii) exposure by ingestion may cause damage resulting from absorbing or swallowing food and water contaminated by pollutants [9]. Typical effects of poor industrial waste disposal are described in the following section.

6.1 Effect on the environment

Poor landfill management and open dumpsites allow emission of greenhouse gases such as carbon (IV) oxide (CO_2) and methane (CH_4) released by decomposing wastes from beverage, petroleum, tobacco, food, or brewery industries; these gases contribute to global warming and cause climate change [38]. Also, hydrogen sulfide (H_2S) released from tannery, beverage, food, and tobacco wastes can cause environmental air pollution [38]. The complex nature of leachate that contains dissolved organic and inorganic toxic substances can cause eutrophication of surface water reducing water quality and amount of dissolved oxygen, which may result in death of aquatic life; leachates can also contaminate groundwater causing shortage of drinkable/usable water or disease and death [3].

6.2 Effect on human health

When landfills and open dumpsites are located near residential settlements due to improper site selection, residents within the vicinity of these sites experience health challenges than people living far away. A landfill that has not been carefully designed and constructed with necessary engineering techniques will contaminate the immediate surrounding due to the release of toxins and bioaerosols [39]. This contamination will lead to health complications such as diarrhea, stomach pain, cholera, flu, asthma, malaria, skin irritation, cough, and tuberculosis [39].

6.3 Effect on wild life

Animals are exposed to toxins from decomposing waste materials disposed at poorly managed landfills or open dumpsites through inhalation and ingestion. These exposure pathways can easily lead to both bioaccumulation and biomagnification of these pollutants in the respiratory track or guts of these animals, which at sublethal concentrations may cause neurological dysfunction, hemorrhaging, and infertility [38]. Lethal concentrations of toxic pollutants in animals may lead to instant death [38].

7. Control measures and Management of Industrial Wastes

The adverse effects of poor disposal of industrial wastes on man, animals, and the environment have resulted in an effort by industries to seek out measures to control and manage wastes generated within their facilities. In addition to the effects, economic and legal (in terms of host community court litigation) implications of waste generation and disposal affect industry-generated revenue. As stated in Ref. [40], the actual costs of generated waste in industry are purchase cost of materials, handling and processing costs, disposal cost, lost revenue, management time and monitoring costs, potential liabilities, and post-disposal segregation. In monitoring and controlling the management of industrial wastes, it is important to carry out pollutant characteristic evaluation so as to examine properties like phytotoxicity, toxicity level (environment and human), persistence, toxic activity, mobility, and bioaccumulation potentials of dioxins in wastes disposed at an engineering landfill facility by government agencies or private bodies so as to determine the effect of their release into the environment [9].

7.1 Waste management hierarchy

In industrial waste management, waste disposal is usually the last resort as industries may lose resources from disposing wastes with reusable or valuable materials often because an industry's waste is another's resource. A framework for deciding method of managing waste in the industry in order to focus on health, safety, environmental protection, sustainability, and cleaner generation of waste is the Waste Management Hierarchy (**Figure 15**) [40]. This hierarchy gives priority according to the waste management methods implemented in the industry and replaces the traditional 3R approach to waste management, namely reduce, reuse, and recycle [41].

The waste management hierarchy points waste management unit to develop a waste management roadmap by first preventing the generation of waste, which is the most preferred management method. When prevention is no longer practicable, waste reduction methods may be implemented through improved manufacturing techniques; after this, the unit would contemplate actions that promote waste reuse. Further, waste recycling options are considered. The next action of the waste management unit would then be energy recovery from the waste, while treatment and disposal are the final actions and the least preferable methods; all of these planning and actionable decisions could be carried out on a six-stage, inverted pyramid shown in **Figure 16** [40].

7.2 Collaborative or Co-operative industrial waste management (CWM)

Unlike municipal wastes, industrial wastes are invaluable or usable raw materials within and across many industries. Waste management techniques, involving reuse, recycle, or energy recovery from wastes, can either be carried out in an industry that generates such wastes or in another industry that has need and capacity to utilize the wastes as resource material or energy source [40]. Waste management carried out by

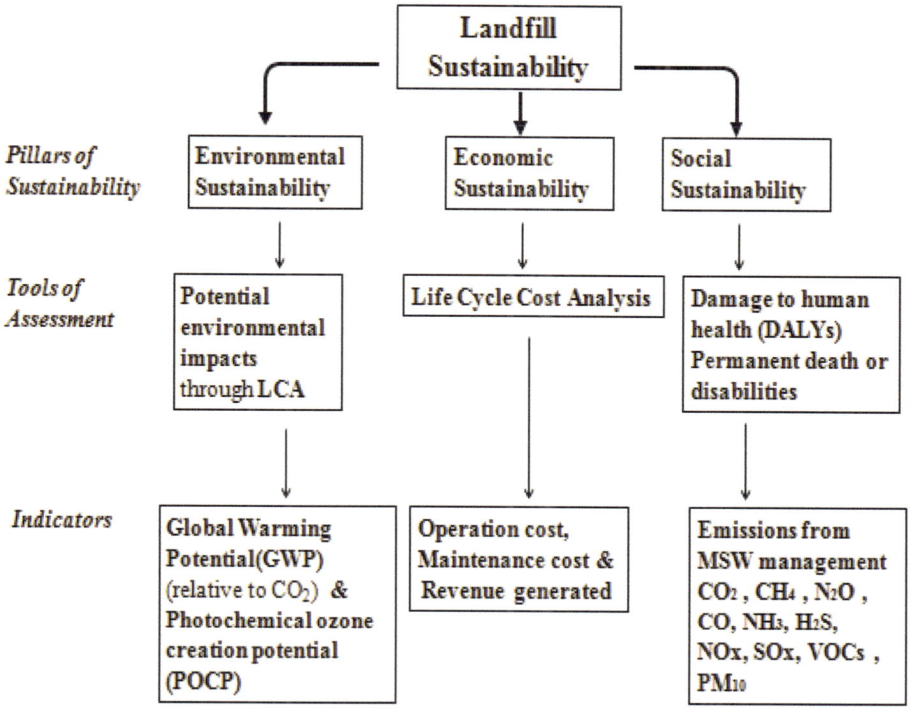

Figure 15.
Landfill sustainability assessment framework [33].

Figure 16.
Industrial waste management hierarchy [41].

a waste-generating industry through another industry is referred to as Collaborative Waste Management (CWM); this collaboration can be direct (**Figure 17**) or through an intermediary (**Figure 18**). CWM is an aspect of circular economy (CE), which is an industrial model that involves production, consumption, reuse, lease, repair, refurbishment, share, and recycling of industrial products in order to enhance retention of raw material value, majorly through circularity [42].

An illustration of collaborative industrial waste management is setup, according to Ref. [43], in Kalundborg, Denmark (**Figure 19**). It consisted of closely located

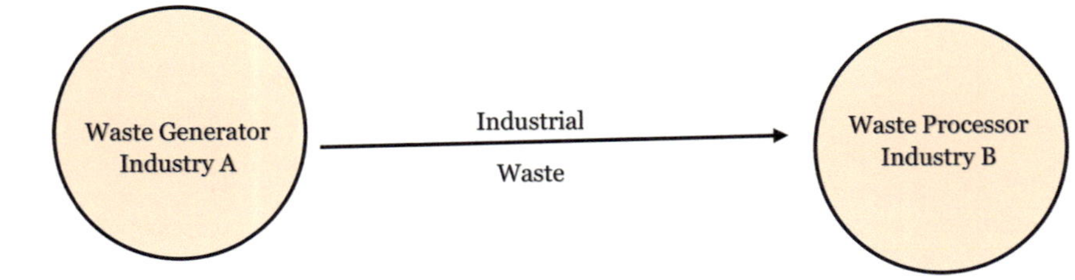

Figure 17.
Direct industrial waste management collaboration [39].

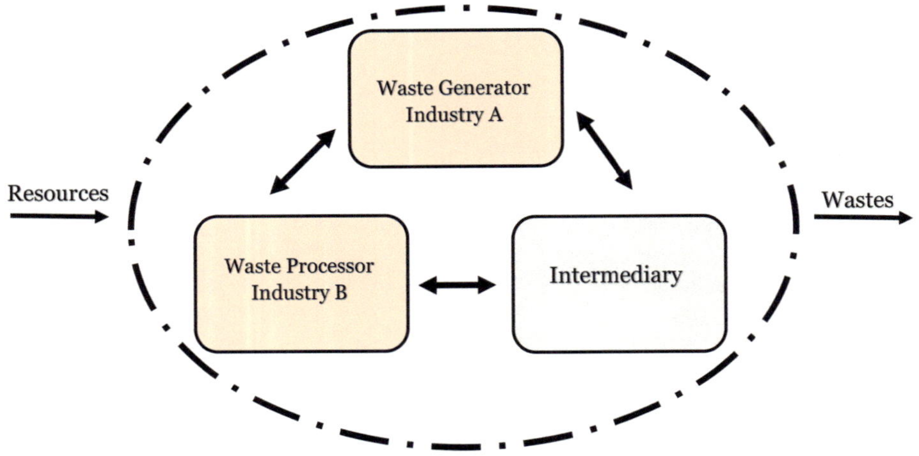

Figure 18.
Intermediary industrial waste management collaboration [39].

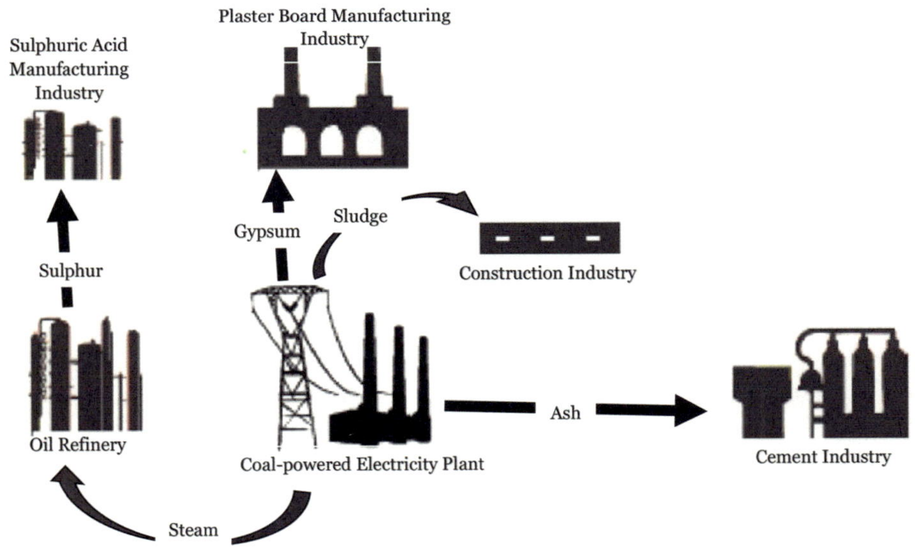

Figure 19.
Industrial waste management collaboration among industries [43].

industries including a cement factory, an oil refinery, a pharmaceutical firm, a coal-powered electricity station, a construction industry, and a plasterboard plant. In this setup, the gypsum, sludge, steam, and ash produced as wastes in the coal-powered plant are used raw materials in plasterboard plant, construction, oil refinery, and cement industries. Also, sulfur produced as waste in the refinery is used as raw material in the sulfuric acid manufacturing plant.

Other instances of CWM approach to industrial waste management can be seen in the use of waste oil and oil-containing materials from petroleum industry as alternative fuel in high-energy demand facilities like cement kilns or the use of spent oils from mining industry as raw material for the production of explosives, such as ammonium nitrate produced by the bulk mining explosive (BME) industry [40]. Also, sawdust, bark, wood scraps, planer shavings, and sanderdust produced in sawmill and woodwork facilities can be recovered and burnt for energy recovery in boiler facilities. There are advantages according to Ref. [40], which can be derived from using CWM method by industries. These advantages can be in terms of economic, social, safety, and legal gains and they include: (i) cost saving on waste disposal, (ii) improved company image with host community, (iii) reduced risk of liability, and (iv) public health and environment benefits.

8. Industrial waste landfill regulations

Regulation, as it relates to industrial waste landfills, refers to the directives or rules made and monitored by a government agency in respect of landfill planning, design, operation, and management. The importance of landfill regulation is to achieve the desired overall environmental health outcome and to reduce the negative impact of landfilling on the environment. The government, which is the major player in making and monitoring adherence to landfill regulations, recognizes that for a new landfill facility, environment protection is achieved through a combination of good planning and a thorough approach to design, operation, and management practices [44]. It is the opinion of Ref. [44] that there is no suitable alternative for selecting an appropriate site and adopting current management techniques to protect environmental integrity of the site. In the case of existing landfills, since the chance to select the best site is long gone, government priority is to ensure that the landfill facility is operated in a way that it minimizes environmental impact, and achieves efficient site remediation. For existing landfills, Peck [44] noted that performance-based approach to management of landfills encourages landfill operators to use their initiative to develop solutions suitable to their landfill operations. According to Peck [44], this approach acknowledges that retrospective design and construction techniques of achieving desired environmental outcomes from a landfill operation could put undue burden of costs on the operators, which, in most cases, is passed onto waste generators *via* disposal fees.

Although the government makes landfill regulations, these environmental rules have to be formulated with wide consultations and bargaining between government agencies and the industry; this is called public private partnership (PPP) for industrial waste management [45]. As stated by Wakiyama [46], industry associations that wield a strong influence on their members are selected to be participants in the formulation of regulatory policy. Government agencies depend on these associations' input to provide information and to measure the costs and feasibility of environmental standards [47]. The PPP allows consideration of industry interests and views in the formulation of regulations so that economic costs and consequences of proposed regulations are accounted before they are put into effect [47]. Once regulations are approved after consultation with industry participation, noncompliance is then monitored and sanctioned by government agencies [47]. When seeking to develop a PPP in industrial waste management projects, the landfill host community's environmental peculiarities are to be taken into account. These peculiarities may include considerations for public health, land ownership, economic and social benefits of the landfill, standards, and accountability, among others [48].

There are essentially two areas of environmental protection regulation for landfills and these are landfill planning and operation. The regulation at planning stage involves obtaining approval for a new landfill or an extension of approval for an existing landfill. Issuance of approval by the government requires that the landfill operator provides a detailed explanation of how the new or existing landfill will directly or indirectly impact the local environment in a document called environmental impact statement (EIS) [44]. The government may achieve operational regulation through issuance of waste licenses whose details are set out in a waste management act such as the Waste Minimization and Management Act of 1996 [49]. A license for operating landfills is required depending on the location of the landfill facility and type of waste received [44]. Typical waste license may include the following criteria:

Location criteria: Landfill location is the single most important determinant of the level of environmental risk posed by a landfill and perfect landfill location is the most efficient environmental management tool used to prevent degradation [44]. This regulation criterion sought to determine early on in the selection process, whether a proposed site is subject to environmental constraint or is of environmental value that it should not be considered as a potential landfill site; in other words, whether the proposed site is environmentally sensitive [44]. For protection of host environment, the location criteria set out steps to be taken when selecting a landfill site, with particular emphasis on carrying out topographic, hydrogeological, geological, and meteorological evaluations to determine the appropriateness of the site [44]. Furthermore, a formal EIS is to be prepared and submitted by the landfill operator if a proposed site will be used for disposal of industrial solid wastes that comprise: about 110,000 tons per annum of nonbiodegradable wastes that are likely to cause flooding or negatively impact drainage, more than 1000 tons per annum of sludge and above 198 tons per annum of other wastes; or the proposed site will be located as follows: within 100 meters of wetlands, coastal dune fields, natural water body; in an area of sodic or saline soil, high permeable or wettable soil, acid sulfate laden soil; within a drinking water catchment or an estuary where sea entrance is intermittently open or on a floodplain; and within 250 meters of a residential area not associated with the landfill site or likely to affect the infrastructure of the neighborhood where the landfill is sited as a result of noise, air pollution, visual impact, traffic, or vermin.

Type of waste and quantity criteria: Waste types disposed at a landfill account for the potential pollutants and likely severity of impact on the environment [44]. For a C&DD landfill that receives inert materials such as construction and demolition wastes that have no potentially hazardous characteristics, environmental risks are generally limited to noise, dust, and sediments, which can readily be controlled thus the criterion here specifies that inert waste landfills must not dispose wastes mixed or contaminated with any other material [44]. Industrial solid waste landfill receiving non-hazardous, degradable waste materials is likely to cause air (odour) and soil (leachate) pollutions as major environmental risks and thus, requires careful management [44]. For these landfills, this criterion provides that solid wastes shall contain less than 200 g/ton of hazardous wastes only and that all solid wastes must have an angle of repose above five degrees (5o) and also must not contain free liquids. Hazardous waste landfills pose the most significant environmental management challenges due to their characteristics and huge potential to cause harm. According to this criterion, hazardous waste landfills must only accept wastes classified as hazardous. These wastes must first be tested, treated, and must meet relevant waste acceptance criteria (WAC) before disposal at the landfill [44].

Environmental risks posed by a landfill can also be gauged by the quantity of waste received at the facility. If a landfill is well located and receives small amount of inert solid wastes, such facility will have minimal impact on the environment, but this is not the same with a facility that receives large amount of waste materials [44]. According to the waste type and quantity criteria, inert waste landfills that receive above 20,000 tons of waste per annum will have to be licensed regardless of their location. Also, 5000 tons per annum capacity solid waste landfills are required to be licensed irrespective of their location. But for hazardous waste landfills, they are required to be licensed regardless of the quantity of wastes received or their location; this criterion is in addition to other special requirements that may be stated by the government regulatory agency in the future regarding disposal of hazardous wastes [44].

According to these regulatory schemes, some facilities will not require licensing due to the relatively small amount of waste they receive and their appropriate or remote location, which indicate that they do not pose a significant environmental risk. The government notwithstanding recognized that if these relatively small facilities are not properly managed, localized environmental damage may result. Thus, it is required that operators make annual notifications to government agencies concerning the quantity and type of waste received, location, and ownership details of the landfill facilities [44].

9. Conclusion

Industrial wastes are unavoidable waste materials generated from industrial manufacturing processes. Many of these wastes materials have characteristics such as toxicity, ignitability, corrosivity, or reactivity, making them not only hazardous but also potential human and environmental health risk factors; although some industrial wastes are nonhazardous. The hazardous nature of industrial wastes has made disposal an important aspect of industrial operations. There are various methods by which industries manage and dispose their wastes but disposal at landfills remains the most practiced due to advantages such as large volume storage, inexpensive operational and maintenance costs, over other disposal methods. Landfilling operations, however, can lead to public health risks if poorly managed or designed due to the formation of by-products like biogas and leachate. Engineering or secure landfill design has features to mitigate contamination from landfill by-products and reduce environmental degradation by preventing leachate from seeping into groundwater and trapping biogas for energy production.

Modern industrial waste management technique focuses on retention of waste material values through reuse by same or other industry as raw material. This technique, called collaborative industrial waste management (CMW), is a subset of circular economy (CE), which is a model employed by large manufacturing industries to harness maximum economic, safety, social, and legal gains from their operations. It is important to operate landfills in a sustainable manner to reduce their negative impact. Also, regulation of landfill operations by the government is necessary to achieve desired environmental health and to mitigate environmental degradation occasioned by landfilling.

Acknowledgements

The authors wish to acknowledge the contributions of Department of Chemical and Polymer Engineering at Lagos State University (LASU) and Department of Chemical and Petroleum Engineering at University of Lagos (UNILAG) for permission to use some of the academic resources consulted for the production of this work.

Conflict of interest

The authors declare no conflict of interest.

Notes/thanks/other declarations

The author wishes to thank IntechOpen for the opportunity to publish their work.

Author details

Olawale Theophilus Ogunwumi[1*] and Lukumon Salami[2]

1 Department of Chemical Engineering, Lagos State University, Lagos, Nigeria

2 Environmental Engineering Unit, Department of Chemical Engineering, Lagos State University, Lagos, Nigeria

*Address all correspondence to: ola.ogunwumi@gmail.com
and oogunwunmi@unilag.com.edu

References

[1] Teku GT. Industrial Waste Management Practices in Addis Ababa: A Case Study of Akaki-Kality Industrial Zone, Ethiopia [Thesis]. Addis Ababa: Addis Ababa University; 2006

[2] Firdissa B, Solomon Y, Soromessa T. Assessment of the status of industrial waste water effluent for selected Industries in Addis Ababa, Ethiopia. Journal of National Science Resource. 2016;**6**(17):1-10

[3] Ogunwumi OT. Kinetic Study of a Lagos Open Dumpsite Leachate Using Activated Sludge Technology [Thesis]. Epe Campus: Lagos State University; 2022

[4] Chinaza GA, Igwe V. Industrial waste management: Brief survey and advice to cottage, small and medium scale Industries in Uganda. International Journal of Advanced Academic Research. 2017;**3**(1):432-488

[5] Jihan Khalid AK. The Impact of Industrial Waste on Human and Natural Resources: A Case Study of Khartoum North Industrial Area [Thesis]. Somalia: Omdurman Ahlia University; 2004

[6] Egyptian Environmental Policy Program. Industrial waste collection and disposal. In: Solid Waste Management Privatization Procedural Manual. Cairo, Egypt: Program Support Unit, Egyptian Environmental Policy Program; 2018. pp. 21-42

[7] Zafar S. Preparing an effective industrial waste management plan. Journal of Bioenergy Consultancy. 2021;**4**:205-209

[8] Millati R, Mohammed DJ, Zadeh T. Agricultural, industrial, municipal and Forest wastes. In: Sustainable Resources Recovery and Zero Waste Approaches. Amsterdam, Netherlands: Elsevier; 2019

[9] Industrial Solid Waste [Internet]. 2010. Available from: https://cpheeo.gov.in [Accessed: August 14, 2022]

[10] Tsai C-H, Shen YH, Tsai W-T. Sustainable material Management of Industrial Hazardous Waste in Taiwan: Case studies in circular economy. Sustainability. 2021;**13**(9410):1-15

[11] United States Environmental Protection Agency. Defining Hazardous Waste: Listed, Characteristic and Mixed Radiological Wastes [Internet]. 2017. Available from: https://epa.gov [Accessed: August 16, 2022].

[12] Dawn Devroom. Hazardous Waste Disposal-IDR Environmental Services [Internet]. 2020. Available from: https://www.blog.idrenvironmental.com. [Accessed: September 24, 2022]

[13] Environmental Health and Safety Unit-Specific Waste Management: Chemical Waste Disposal and Management [Internet]. 2018. Available from: https://www.ehs.utoronto.ca [Accessed: August 9, 2022]

[14] United States Environmental Protection Agency-Radioactive Waste: About Radioactive Waste Disposal and Management [Internet]. 2017. Available from: https://epa.gov [Accessed: September 27, 2022]

[15] Hari D. Industrial Waste and Waste Management [Thesis]. Shamshabad, New Delhi: Vardhaman College of Engineering; 2018

[16] Trends in Solid Waste Management-What a Waste 2.0-Global Snapshot of Solid Waste Management to 2050 [Internet]. 2022. Available from: https://www.theworldbankgroup.org [Accessed: August 22, 2022]

[17] Industrial Waste. Industrial Waste Factsheet: Safe Drinking Water Foundation [Internet]. 2016. Available from: https://www.safewater.org [Accessed: September 27, 2022]

[18] Frost and Sullivan. The Global Industrial Waste Recycling. Santa Clara, California: AMP Services Market; 2012

[19] Gourav S, Naveen BP, Malik RK. Industrial Hazardous Waste Management [Thesis]. Haryana, Gurugram: Amity School of Engineering and Technology; 2017

[20] Veronica G, Eva P, Riitta K. Waste Minimization in the Chemical Industry: From Theory to Practice [Thesis]. Finland: University of Oulu; 2004

[21] Xiaoyuan Z, Ping G, Yu L. Decontamination of radioactive wastewater: State of the art and challenges forward. Chemosphere. 2018;**10**(029):34-39. DOI: 10.1016/j.chemosphere.2018.10.029

[22] David Fahrion. Ways by which Industrial Recycling Can Benefit Your Business: Waste Control Blog [Internet]. 2014. Available from: https://www.wasteconrolinc.com. [Accessed: August 25, 2022]

[23] Parliament of Australia-Waste Management and Recycling in Australia [Internet]. 2016. Available from: https://www.aph.gov.au [Accessed: September 20, 2022]

[24] Mullen L. What Is Industrial Composting? [Thesis]. USA: Environmental Centre, University of Colorado; 2020

[25] Merry's System-United Nations Industrial Development Organization. Food Waste Composting and Recreating Recycling Loop: An Effective Solution to Food Wastes by a Community-based Composting System [Internet]. 2020. Available from: https://www.unido.or.jp [Accessed: September 20, 2022]

[26] Bebar L et al. Secondary combustion chamber with inbuilt heat transfer: Thermal model for improved waste-to-energy systems. Modeling chemical. Engineering Transactions. 2010;**21**(33):859-864. DOI: 103303/CET1021144

[27] Carlo Trozzi. Industrial Waste Incineration: EMEP/EEA Emission Inventory Guide Book [Internet]. 2009. Available from: https://www.eea.europa.eu [Accessed: September 27, 2022]

[28] Andrew K. An Overview of Incineration and EFW Technology as Applied to Management of Municipal Solid Waste [Thesis]. USA: University of Western Ontario; 2005

[29] Andreja G, Aleksander P. Perspectives on biological treatment of sanitary landfill leachate. In: Waste Treatment Engineering. London, UK: InTech; 2015. DOI: 10.5772/60924

[30] Conservative Energy Future-What is Sanitary Landfill and Difference between Sanitary Landfill and Open Dumping [Internet]. 2013. Available from: https://www.conserve-energy-future.com [Accessed: September 27, 2022]

[31] Rinkesh W. What is solid waste management? - sources and methods of solid waste management. Journal of Conservative Energy Future. 2018;**2**:809-819

[32] Scharff H. The role of sustainable landfill in future waste management. In: NV Afvalzorg Holdings, 1566ZG Assendelf. Netherlands: Afvalzorg Holding; 2019. pp. 1-8

[33] Sivakumar et al. Assessment of Landfill Sustainability. Researchgate. Available from: https://www.researchgate.net/publication/310740034

[34] Kerry IH, Ann DC, Joe EH. Landfill Types and Liner Systems: Extension Factsheet [Thesis]. Columbus: The Ohio State University; 2014

[35] Yashar R et al. Landfill site selection using multi-criteria decision making: Influential factors for comparing locations. Journal of Environmental Sciences National Centre for Biotechnology Information. 2020;**93**:170-184

[36] Vesilind PA, Worrell W, Reinhart R. Industrial Solid Wastes. Journal of Solid Waste Engineering Management. Brooks Cole. 2002;**1**:201-208

[37] Awuchi et al. Industrial waste management treatment and health issues: Wastewater, solid and electronic wastes. European Academic Research. 2020;**8**(2):1081-1119

[38] Awuchi and Awuchi. Physiological effects of plastic wastes on the endocrine system (Bisphenol a, phthalates, Bisphenol S, PBDEs, TBBPAs). International Journal of Bioinformatics and Computational Biological. 2019;**4**(2):11-29

[39] Ofoezie EI, Sonibare J. Health and environmental consequences of industrial wastes and toxic chemicals: A review article. Research Gate. 2004;**1**:161-175

[40] Njoku PO, Edokpayi JN, Odiyo JO. Health and environmental risks of residents living close to a landfill: Case study of Thohoyandou landfill. Limpopo Province, South Africa. International Journal of Environmental Research and Public Health. 2019;**16**(2125):1-24

[41] Gayani K, Rathnayake R. Industrial hazardous waste management: Avenues for collaborations. In: Proceedings of the 7th FARU International Research Symposium. Berlin, Germany: Researchgate; 2013

[42] Axil Integrated Services-What is a Waste Management Hierarchy? [Internet]. 2012. Available from: https://www.axil-is.com [Accessed: August 21, 2022]

[43] Salmenpera H, Saikku L. Critical factors for enhancing the circular economy in waste management. Journal of Cleaner Production. 2021;**280**(1):1-21

[44] Peck SW. When is an eco-Industrial Park not an eco-Industrial Park? Journal of Industrial Ecology. 2012;**5**(3):3-5. DOI: 10.1162/10881980176007413

[45] Environmental Guidelines: Solid Waste Landfills [Internet]. 1996. Available from: https://www.epa.gov.au [Accessed: August 21, 2022]

[46] Wakiyama T. The Implementation and Effectiveness of MITI's Administrative Guidance in Corporate Government-Industry Relations: Western Europe, United States and Japan. Oxford: Claredon Press; 1987

[47] Yasunori S. Structural, Plitical Bargains: Government, Gyokai and Markets. Ithana, New York: Cornell University Press; 1993

[48] Mitsutsune Y. Planning and implementing environmental policy and the role of industry: The relationship between government and industry-the situation in Japan. In: Conference on Lesson of Japanese and United States Environmental Policy for Industrialized and Developing Countries. Washington, D.C.: Centre for Global Change, University of Maryland; 1994

[49] Waste Minimization and Management Act [Internet]. 1996. Available from: https://www.revisedacts.lawreform.ie [Accessed: September 27, 2022]

Section 2

Technologies and Practice

Chapter 5

Efficient Treatment of Municipal Solid Waste in Incinerators for Energy Production

Terrence Wenga

Abstract

MSW generation has increased drastically throughout the world surpassing the ability of municipalities to handle it. Treating waste in incinerators with energy recovery have been opted as an environmentally preferred method of waste management. However, waste incineration result in inefficient energy generation. The objective of this chapter is to provide a summary of issues leading to inefficient treatment of MSW and the potentials for improving it. High-temperature corrosion and ash-deposition on heat exchange surfaces are the major causes of inefficiency during waste incineration. Optimizing the operating conditions during incineration reduces the deterrent corrosion and ash deposition problems. The operating conditions can be optimized by conducting a kinetic modelling which identifies the conditions that reduces corrosion rate. These conditions are moisture content ~10 vol.% and SO_2 ~250 ppm. Also, use of ecotubes and sergher-boiler prisms ensures high turbulence and mixing within the boiler which reduces the ash problems, thereby improving the incineration efficiency. Sorting of MSW using max Al robotic sorter and removal of alkali chlorides in waste through the use of sink-float process, centrifuge and hydrocyclone separation technologies lowers chlorine load hence lowering the severe ash problems and proves to be beneficial in improving the efficiency of treating MSW in incinerators.

Keywords: MSW, incineration, high-temperature corrosion, ash deposition, combustion efficiency

1. Introduction

The growing population and economic development have resulted in increased municipal solid waste (MSW) generation which surpasses the current ability of the municipal authorities to handle it [1, 2]. Landfilling is the utmost preferred method [3] but the least considered method for waste disposal and management owing to the release of methane gas and discharge of the leachate into the ground water, thereby polluting it and rendering it unsafe for domestic purposes [4]. The emissions of methane and other greenhouse gases (GHGs) have induced and inflicted global warming which in turn affected the sustainable development of countries, particularly in the developing world. As a result, incineration of MSW in incinerators for energy recovery has been considered as the most preferred method for sustainable waste management, and safe disposal of waste, because of the plusses of quick in mass and volume reduction by ~70% and ~ 90%, respectively, electricity and heat energy recovery, as well as complete disinfection [1, 5].

Currently, around 220 million tons of MSW are treated globally, in over 800 waste-to-energy (WtE) incineration plants [6]. Data showed that the total energy produced globally from municipal wastes in 2010 was 41.743 GWh, while in 2015 it had increased to 62.507 GWh [7]. This increase in energy generation indicates that combustion of MSW in WtE incinerators has drawn increasing attention. Aside from generating energy, WtE incineration plants have the benefit of lowering GHG CO_2 emissions per unit of coal substituted, since waste fractions are largely biogenic approximately 70% are combustible organics materials [8–10]. In 2007, the worldwide impact of waste combustion in WtE incineration facilities on climate change was estimated at 40 million tons CO_2-eq, compared to 700 million tons CO_2-eq obtained from landfilling of waste, and in 2015 at 60 million tons CO_2-eq compared to 800 million tons CO_2 for landfilling [9]. With the policy of diverting waste from landfills toward incineration, it is expected to reduce approximately 92 million tons CO_2 equivalence per year by 2030 which is approximately 8% of 1137 million tons CO_2-eq, that is predicted to be reduced by 2030 [1, 11].

Nevertheless, wastes contains elevated chlorine concentration (0.45–1.00 wt. %) [10, 11] and salts of alkali metals especially Na and K, its incineration leads to grave fouling, ash deposition and high temperature corrosion of the boiler tube metals [11–16]. It has been estimated that chlorine-induced corrosion of the boiler tube metals reduces the electricity generation efficiency of WtE incineration plants by approximately 0.5–1.5% [17]. This occurrence forces the incineration plants to undergo an unplanned shut down for maintenance, causing loss in the treatment of wastes and reduced energy and electricity production. Generally, about 75% of the budget for the planned shutdown are consumed in the maintenance of the degraded boiler tubes in this unplanned downtime, resulting in economic loss [1].

The chemical reactions that occur in the gas phase between chlorine, potassium, and sulfur determine the species that reach and react with the boiler tube surfaces [18]. These gas phase and surface reactions are complicated and cannot be determined by experiments due to the inability of online measurements. Thermodynamic equilibrium calculations are normally used to predict the reaction products. Nevertheless, such predictions give only conditions that are thermodynamically stable whereas in actual systems the local kinetics controls the corrosion process [19]. The main steps in corrosion include the diffusion of gaseous species from the combustion environment to the surface, adsorption of the reactants onto the metal surface, reaction with the surface, desorption of the volatile products from the surface and diffusion of the products back to the combustion environment. The diffusion of gases is driven by the concentration gradient. However, convection, temperature gradients, and pressure gradients lead to the deviations of this flow [20]. The surface reactions involve the formation of intermediates which then reacts with further gaseous species from the flue gas [19]. The intermediates formed in both gas and surface reactions are difficulty to detect experimentally [21–23]. Therefore, it is necessary to study corrosion by combining theoretical calculations and experiments. Kinetic modeling reveals the fundamental nature of the reaction mechanisms [21]. The corrosion problem and other fireside challenges leads low MSW treatment efficiency.

Literature on hindrances and options to improve MSW incineration efficiency are scarce. Ma et al. [1] described the challenges occurred during incineration, however, that paper is not specifically on the options to improve waste incineration efficiency. Kumar and Samadder [12] reviewed technological options for effective energy recovery and the challenges faced in both developing and developed countries. The paper did not focus on methods to improve treatment efficiency. Prajapati et al. [13] summarized the recent technological advancements focusing on minimization of waste content and electricity generation focusing on biogas production. Varjani et al. [2] gave an overview of the existing sustainable MSW management technologies together with their limitations. While some attempts have conducted on the sustainable treatment of MSW, no study has elaborated on the option to improve the treatment efficiency. Understanding the details will assist in the design of the incinerator with less fireside challenges during combustion of MSW hence the treatment of MSW becomes sustainable. The purposes of the chapter was to shed light on the challenges that causes inefficient treatment of waste in WtE incinerators as well as the potentials for improving the efficient treatment of waste in incinerators. The results from this research will help incineration plant operators to implement and improve the efficiency of treating MSW, thereby becoming economically viable and competitive.

2. Methodology

The approach employed for retrieving the literature used in this chapter is summarized in **Figure 1**. Briefly, the literature was obtained from published articles found on academic databases which are Clarivate's Web of Science®, ScienceDirect®, Researchgate®, Google Scholar®, and Scopus®. A Boolean search method employing the 'AND'/'OR' were utilized in searching for literature upon entering the

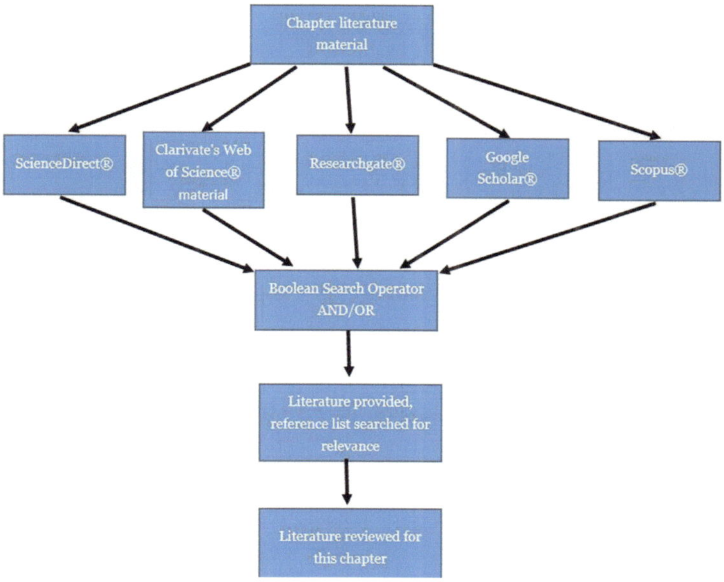

Figure 1.
Flow diagram for content retrieval.

keywords. For example, the keywords used are MSW, incineration, high-temperature corrosion, ash deposition, combustion efficiency and kinetic modeling. A thorough cross search from the reference listed in the relevant articles were conducted for their significance to the current chapter.

3. MSW as a source of energy

Solid waste is composed of a range of constituents, which are not all recyclable/ reusable [14]. Due to ever increasing quantities of solid waste generated throughout the world and the fact that approximately 70% of it is organic, it is regarded as a huge source of renewable energy [14, 15]. In addition, MSW is abundant, free and available everywhere, recovering energy from it has gained much attention, globally, with the developed countries carrying the process to the full capacity of the technology [16]. MSW presents various drawbacks during incineration. These include low energy content approximately 10–13 MMBTU/ton, an amount which is far less than sub-bituminous coal which has about 17–21 MMBTU/ton [17]. In addition, solid waste has a high moisture content approximately >50%. Recovering energy from waste simply means that huge supply of energy is required for igniting and drying the waste for combustion to take place. Its ultimate composition is extremely heterogeneous making combustion very difficult and unstable and different municipalities can use completely different waste pre-sorting approaches. Despite the heterogeneity nature of its ultimate composition, its elemental composition is also diverse. Levels of chlorine, sodium, potassium, sulfur, nitrogen, heavy metals (e.g., zinc, lead, cadmium and others) and ash content in waste are greater than those in lingo-cellulosic feedstock [4]. When combusted, these elements react with oxygen or react among themselves producing pollutants e.g., oxides, chlorides, dioxins and furans. Incineration plant can be affected by these pollutants, it need frequent clean-up [17]. Apart from that, MSW is occurs everywhere on land and collecting it to one point is a challenge and costly [14]. In general, densely populated areas produce substantial amounts of MSW and incineration plants can be installed in such areas. Locations which are less populated require new, smaller-scale technologies that need small amounts of waste as feedstock [14, 17].

3.1 Energy recovery from waste incineration

MSW is a complex assortment of combustible materials such as food waste, paper, yard trimmings, plastic, as well as sludge from wastewater treatment and incombustible materials including metals, glass, rags, construction and demolition waste which are intermingled [16]. **Table 1** shows the ultimate analysis of the MSW fractions. The most preferred approach of recovering energy from MSW is incineration. If the amount of energy that can be produced from the recovery process is greater than R1 formula, then waste energy recovery is viable but if the amount of energy is less than R1 formula then waste should be disposed of in landfills [18]. Incineration technology has the advantages of killing pathogens as well as fast-volume and mass reduction of waste by approximately 90% and 70%, respectively, with the residual waste going to the landfill [19]. Because incineration plant is generally installed close to the source of waste, it also reduces the distance of hauling municipal wastes to the plant. Although some authors [16, 20] point that, these advantages are equipoised by emissions

Component	C	H	O	N	Cl	S	K	Na	Zn	Pb	Ash	Moisture	Ref
	wt.%					mg/kg					wt.%		
Food items													
(Mixed) Food waste	48.0	6.4	37.6	2.6		4000					5.0		[18]
Meat	42.51	7.03	30.20	13.08	2.74	8100	0.78	1.6	0.018		4.9	75	[19]
Plastics													
PVC	36.4	4.43		<0.1	6.3[a]	525	70	267	140	3530	4.98	0.19	[7, 20]
Paper													
Cardboard	43.0	5.9	44.8	0.3		2000					5.0		[18]
Glossy paper	27.4	3.5	34.5	0.07	0.06	200	139	1415	3.97	1.31	34.5		[19]
Waxed cartons	59.2	9.3	30.1	0.1	0.1								
Magazines	32.9	5.0	38.6	0.3		1000					23.3		[18]
Wood													
Wood waste	50.4	6.06		0.16	<0.01	419	157	463	50.5	26.9	1.81	10.44	[21]
Yard wastes	41.54	4.79	31.91	0.85	0.30	0.24	6000	2000	—	—	20.37		[19]
Bark	53.7	5.9	39.9	0.43	0.02	0.05	2100	0.04			4.8	52.3	[22]
Others													
Shoes	58.8	7.27		1.92	3.20	8610	533	892	6840	243	15.5	2.58	[7]
Leather	54.9	5.1	19.2	14.1	0.8	14,000					5.25	13.3	[23]
Batteries	—	—	—	—	1.55	5300	4900	11,800	1787	830			[24]
Shredder waste	17.3	2.15		0.59	0.33	4890	4630	11,900	22,000	2810	72.8	2.05	[21]
Sewage sludge	53.5	6.4	36.1	2.63	0.12	12,600	3600	3800			28.6	75.6	[22]
Tires	84.39	7.13	2.19	0.24	0.15	12,400	—	—	65	1.53	4.81	0.62	[19]
MSW	39.5	5	26.5	1.5	0.7	3000	2500	2500	300	300	26.5	25	[19]

[a] *Cited from Ref. [20].*

Table 1.
Ultimate analysis of MSW fractions.

of oxides of sulfur and carbon, heavy metals, particulates, and dioxins where it is estimated that each ton of MSW incinerated produces approximately 15–40 kg of hazardous waste, others [1, 21] argue that recent incineration technologies are equipped with sophisticated air pollution cleaning system where pollutants emitted are within the standards [16]. Nevertheless, incineration technology is still claimed to produce dioxins and furans into the environment which are toxic but indeed there is still no report that is available of anyone who have been affected by dioxin from current incinerators.

The most used facilities for the incineration of waste are modular systems, mass burn as well as refuse derived fuel systems. In the mass burn technologies, waste is converted to energy through the mass combustion process. The waste is incinerated as received with or without prior sorting before entering the burning combustion chamber [22]. This technology burn waste in an incineration chamber supplied with surplus air which stimulates the complete mixing of waste with air and causes turbulence with the chamber to ensure that there is homogenous mixing for complete combustion. This homogenous mixing is vital owing to the heterogeneity of solid waste. Mass-burn technology burns waste on a sloping, moving grate that vibrates to shake the MSW mixing it with air [23]. In the modular Systems, waste is combusted without being processed. This technology is different from mass burning because they are smaller and movable from one place to another. In refuse derived fuel (RDF) technologies, MSW is shredded by the mechanical methods. This removes incombustible materials from the combustible fractions thus increasing the heating content of the waste which can be used as fuel in RDF furnace [22].

The number of waste to energy incinerators has amplified from an approximately 200 to over 800 plants worldwide now [24]. With a mean calorific value of 10 MJ/kg from the MSW, it is estimated that approximately 500 kWhe/ton waste of energy is produced from the combustion of waste. This amount of energy can be transformed to usable energy forms such as heat and electricity. The energy generated maybe increased by elevating the steam parameters i.e., steam temperature and steam pressure. However, this is hindered by (i) high-temperature corrosion of the boiler superheater tubes, (ii) fouling and ash deposition on tubes which then reduces the heat transfer to the steam from the flue gas, and (iii) fluctuation in steam temperature [24, 25]. Controlling these challenges results in the sustainable treatment of waste.

4. Problems faced during incineration

Incineration is the most preferred method for the treatment of waste and it releases energy which can be utilized for heat and electricity generation. Due to the high chloride, alkali and sulphate content in the MSW, the technology suffers grave high temperature corrosion, ash deposition, and fluctuation in steam temperature thereby limiting efficient utilization or treatment of waste [24].

4.1 High temperature corrosion

Corrosion that occur at elevated temperatures in waste to energy technologies has been extensively investigated and the attack is caused by large amounts of chlorides and sulphates that appear in the flue gas of the combustion gas. Chloride are more dominant in inducing the corrosion attack than sulphate in waste to energy plants due to the elevated chlorine concentrations in MSW. HCl and Cl_2 generated are emitted into the flue gas as the combustion proceeds, where HCl occurs in bulk gas containing moisture, while Cl_2 occur in a dry environment, and may also result from the decomposition of HCl [26]. The chlorine-induced corrosion mechanism is generally understood as the active oxidation process. The main corrosion reactions taking place in the active oxidation mechanism, are illustrated in reactions (1)–(5) and in our review paper [1]:

$$2HCl(g)+1/2O_2(g)=Cl_2(g)+H_2O(g) \tag{1}$$

$$M(s)+Cl_2(g)=MCl_2(s) \tag{2}$$

$$MCl_2(s)=MCl_2(g) \tag{3}$$

$$2MCl_2(g)+3/2O_2(g)=M_2O_3(s)+2Cl_2(g) \tag{4}$$

$$3FeCl_2(g)+2O_2(g)=Fe_3O_4(s)+3Cl_2(g) \tag{5}$$

Where M = Fe, Cr or Ni.

When the conditions inside the boiler have plenty supply of oxygen and temperatures is less than 600°C, HCl(g) become oxidized at the deposit/gas interface, giving Cl_2 gas. However, when the temperature is greater than 600°C while moisture is present, HCl generation is promoted [27]. Chlorine that is produced volatises and evaporate through the pores and cracks to the scale/metal interface, where oxygen partial pressure is low, and reacts with the metal elements producing solid metal chlorides. Due to high volatility, metal chlorides evaporate continuously and diffuse to the gas-oxide boundary and react with oxygen thereby converted to oxides forming a porous oxide layer that cannot prevent further inward diffusion of chlorine gas to the metal substrate for corrosion attack. The chlorine released diffuses back to the metal surface; therefore, a cycle that, with little or no net depletion of chlorine, provides a continuous removal of metals away from the metal surface, toward regions with higher oxygen partial pressure, where oxides are formed [28].

While the active oxidation process has credibly received experimental support [28–32], some problems with the corrosion mechanism appear to have been ignored which led to the difficulties in actually finding the solution to the corrosion problem. The problems are that why the metal/scale interface has low oxygen partial pressure, but high for chlorine? Does this imply that oxygen is prevented to diffuse across the scale, but chlorine, which has a large molecular size, is allowed? Li et al. [33], suggested that the reaction of oxygen in the oxidation of alloy elements, particularly Cr, at the metal/scale interface, decrease the partial pressure of oxygen, while Cl_2 continue diffusing into the interface leading to increased Cl_2 partial pressure, at the metal/scale interface which in turn react with tube metals.

4.2 Kinetic modeling of corrosion

High temperature corrosion can be studied either by experiments (lab-scale, pilot-scale, and commercial-scale studies) or by theoretical studies which include kinetic modeling and also thermodynamic equilibrium as well as CFD modeling. The chemical reactions that occur in the gas phase between corrosive gases such as chorine,

Element	Lowest case scenario (wt.%)	Worst case scenario (wt.%)
K	0.16	0.2
Na	0.04	2.03
Cl	0.01	1.13
S	0.08	0.08
Zn	0.012	0.056
Pb	0.045	0.15

Data was taken from various references [35–37].

Table 2.
Composition of MSW.

potassium and sulfur during combustion of MSW determine the species that reaches and reacts with the boiler tube surfaces [34]. **Table 2** shows the concentrations of elements that are found in MSW and that participate in corrosion process for both the lowest and the worst case scenarios. These gas phase and surface reactions are complicated and cannot be identified by experiments due to the inability of online measurements. Thermodynamic equilibrium calculations are normally used to predict the reaction products. Nevertheless, such predictions give products under conditions that are thermodynamically stable whereas in actual systems the local kinetics controls the corrosion process [38]. The main steps in corrosion include the diffusion of gaseous species from the combustion environment to the metal surface, adsorption of the reactants onto the metal surface, reaction with the surface, desorption of the volatile products from the surface, and diffusion of the products back to the combustion environment. The diffusion of gases is driven by the concentration gradient. However, convection, temperature, and pressure gradients lead to the deviations of this flow [39]. The surface reactions involve the formation of intermediates products which further reacts with either gaseous species from the flue gas (Eley–Rideal mechanism) or with other adsorbed intermediates (Langmuir–Hinshelwood mechanism) [38]. Since experiments have the inability of online detection, theoretical calculations provide new and useful information regarding the corrosion phenomenon and optimisation of the combustion environment for reduced corrosion rates [39].

Ma et al. [40] developed a kinetic model for the surface reactions between gaseous K, Cl, and S species with pure Fe metal during corrosion. The model was employed to explore the effect of KCl on the corrosion of pure iron metal. The amounts of KCl used resulted in different K/S ratios. The authors observed that increasing KCl amounts, which increases K/S ratio, accelerated the corrosion of Fe metal as illustrated in **Figure 2**.

Chen et al. [41] further employed kinetic modeling to investigate the operating conditions that reduces the corrosion rate on the boiler tube metals. They investigated concentration of sulfur dioxide, moisture content, hydrochloric acid concentration and influence of steam pressure and temperature. The authors found that the concentration of sulfur plays a major role in reducing the corrosion effect. The concentration of SO2 between 0 and 270 ppm resulted in less effect on corrosion impacts. This was ascribed to small amounts of sulfur in the combustion system which could not induce any effect. However, the concentration between 270 and 500 ppm resulted in a reduced corrosion rate of the boiler tube metal. The author put forward that the concentration of sulfur was sufficient to transform the corrosive potassium chloride to less corrosive potassium sulphates. The sulfur contents increased the reaction

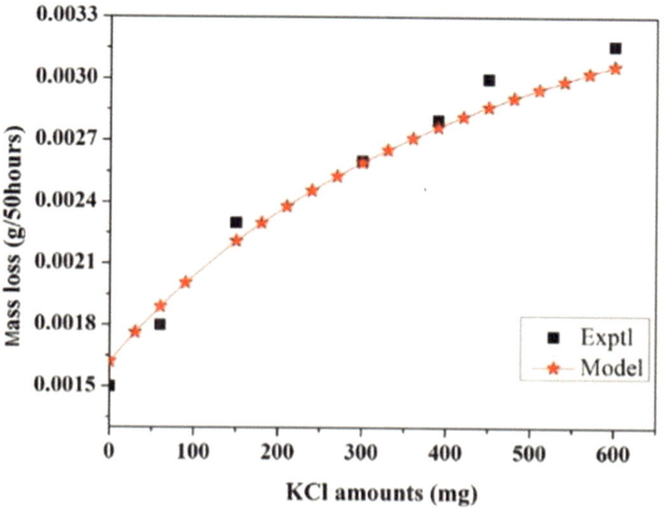

Figure 2.
Influence of KCl on corrosion of iron metal [40].

activity for KOH + SO_3(+M) = $KHSO_4$(+M), which also has a reverse reaction. This on-set a reaction loop where potassium is sulfated to $KHSO_4$, decompose back to KOH and sulfated again. The continuation of the cycle eventually leads to sulphation of all the available K due to K/H exchange between KOH and $KHSO_4$ to K_2SO_4. Sulfate formed prevents reactions between Cl and Fe, thus corrosion decreases. A concentration above 500 ppm was observed exacerbates the corrosion impact. Chen et al. [41] explained the mechanism by saying that when FeS is formed, and because there are cation vacancies in FeS structure, there is high diffusion rate of Fe ions through FeS facilitated by iron concentration gradient established by more stable iron oxide formed at the exterior surfaces leading to increased mass loss [42]. The influence of moisture was investigated and it was observed that moisture content ~10vol. % resulted in a reduced the corrosion rate. The inhibitory effect of H_2O vapor on corrosion rate was due to the facilitation of the formation of K_2SO_4 [41]. The corrosion mechanism of superheater made up of pure iron is shown in **Figure 3**.

4.3 Ash deposition

During combustion of waste, ash particles of various size migrate from the combustion bed to the superheater surfaces where they form ash deposits. Ash build-up during is divided into two mechanisms which are influenced by the flue gas temperature in the boiler. The first one is via solidified slag formation and the second one is the powdered ash depositions. The solidified slag deposition is formed between 1070 and 1320 K and typically has high contents of Fe_2O_3 and sulfates and low contents of SiO_2 and Al_2O_3. The powdered ash deposition occurs below 1070 K and contains more than 50% SiO_2 and over 20% Al_2O_3 [43].

After reaching the boiler tube surfaces, the incident gases heterogeneously react with the tube surface, resulting in the collection of mass of the ash deposit. Chemical reaction cause the vapor pressure of a species to be zero on the deposit surface, hence keeping a concentration gradient with the bulk gas which continuously allows the diffusion processes and reactions to occur [44]. The most key reactions that facilitates

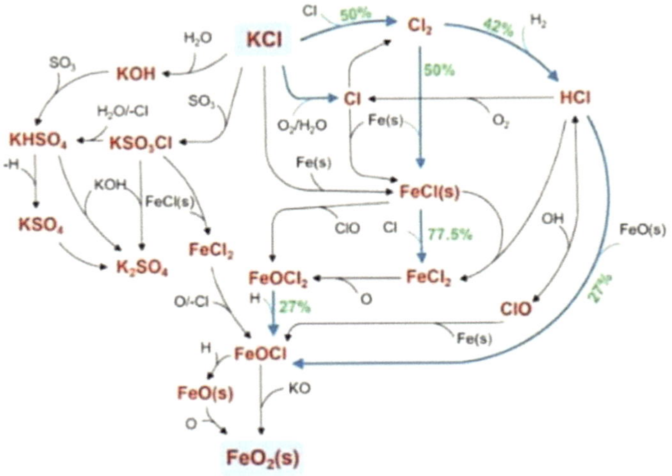

Figure 3.
Corrosion mechanism of pure iron metal.

the ash deposition and growth are alkali absorption, sulphation, oxidation or reduction, carbonation and creation of eutectics due to reaction of K, Na, Fe, Ca, Si, and Al [1]. The main species causing sulphation are compounds of alkali metals, such as potassium and sodium hydroxides and chlorides. This was evidenced by Hansen et al. [45], who observed a potassium sulphate layer in the deposits, which were exposed to the waste combustion conditions for over a period of 1 year. On the contrary, the probes which were exposed for a short period of times contained only KCl without K_2SO_4. Therefore, it is concluded that KCl is the species that initiate the ash deposition by forming eutectic mixtures with $FeCl_2$ at 355°C, with K_2SO_4 at 690°C, and also with Fe_xO_y. The eutectic mixture then sticks on the surfaces of the boiler tube then followed by reaction with sulfur-containing species, such as gaseous SO_2 or SO_3 forming potassium sulphate. Deposition forms when ash accumulates on the heating surface at temperatures lower than the ash fusion point. Detailed ash deposition mechanism can be found in the [46]. Once the ash deposits are formed, they induce solid state high temperature corrosion of heat exchange surfaces, reduce heat exchange from the flue gas to the heated medium and, hence, inhibit heat transfer in the boiler and reduce the boiler efficiency. Once the boiler efficiency is reduced, waste treatment rates becomes compromised.

5. Options to improve MSW treatment efficiency

Waste-to-energy incineration represent an existing technology for efficiently dealing with the most problematic waste issue worldwide. To achieve optimal resource recovery, production of heat and electricity, it require prior separation of recyclables. However, high temperature corrosion, ash deposition, fluctuation in steam temperature, and high operating costs make the treatment of MSW in incinerators a challenging. Several opportunities are available to improve the treatment of wastes in incineration facilities.

5.1 Waste pre-processing

Pre-sorting of waste enables the collection of some waste fractions to be recycled, and those that cannot be recycled be transformed to electricity, as well as production of value added products. Research breakthrough comprises of characterization approaches for high-precision sorting such as the development of max AI recycling robot. The system can work safely alongside people in the system, which allows the Max-AI robotic sorter to be placed into existing material recovery facilities or newly built waste to energy facilities where waste as received from the society is sorted before either recycle, reuse or converted to energy.

5.2 Removal of dissolved pollutants

Quality control measures and pre-treatment processes to remove contaminants can improve the treatment of waste. This encompasses removal of dissolved pollutants from waste such as chlorine, alkali metals and sulfur which causes high temperature corrosion and ash deposition.

To sustainably and efficiently treat MSW in incinerators, minimum chlorine-induced adverse effects should be observed. Therefore, it is crucial to control the dissolved pollutants prior to the burning process such as the removing the alkali chlorides and separating plastic, than attempt to minimize their damage by the chemical dechlorination and catalytic separation. Since all the incineration hindrances are mainly due to the presence of chlorine, much research have focused on understanding the relationships between chlorine in the waste and the HCl formation. An investigation was conducted to determine the amount of HCl that is emitted from PVC and a mixture of waste without the PVC-fraction. It was observed that HCl-emission decreased by 40% to 1.7 mg Cl/g wet MSW [47]. Another authors Delay et al. [48] showed that the decrease in plastics, especially PVC in the MSW result in decreased boiler corrosion as well as decreased heavy metal emissions. At present, the techniques used to separate plastics, are grounded on density. In consideration of its accuracy, three advanced separation processes: the sink-float process, centrifuge and hydrocyclone separation are used, in order to find out how each process would distribute chlorine contained in plastic waste, to the overflow and underflow. It is concluded that with the current sorting technology, the 13% mixed waste plastic go to incinerators, 35% to recycling, and 52% to landfills. With the sink-float separation technology, 47.5% chlorine-poor plastic (<0.5%) goes to incineration plants as alternative fuel, and 16.9% (ca 4.7% Cl content) landfilling.

For the inorganic salt in the waste streams, a water-washing process can be a simple but practical method for inorganic salts removal. Numerous studies on waste and biomass washing have revealed that a large portion of chlorine could be released by elution tests. Jensen et al. [49] reported that approximately 90% of the salts can be removed from biomass char, within 20 min by water washing. Chen and Pagano [50], observed that application of a high temperature (93.7°C) leaching technique on a high chlorine coal (0.5 wt.%) result in chlorine decrease to about 0.2 wt.%. In MSW approximately 85% of easily water-soluble chlorine was leached out in the first washing procedure by distilled water [51]. The limitation of the water extraction method is on increasing the total costs of the operations as well as disposal of the effluent.

5.3 Optimisation of combustion environment

Optimization of combustion conditions in incineration plant had recently emerged as a method to reduce the corrosion impacts in waste to energy plants. Among the key methods, installations of the Segher Boiler Prisms in incineration facilities [47] has been utilized as a measure reduce corrosion in the boiler. The technology consist of a prism-shaped dynamic secondary air mixer, that is inserted at the transition of the combustion chamber and is cooled by flowing water, lined by refractory materials, and has the natural circulation system. The prism splits the flue gas into two channels where each is supplied by secondary air injection. Prism ensures homogenous injection of secondary air via multiple nozzles that are on the prism sides and boiler walls. This supply ensures that there is high turbulence and mixing of waste with excess air which prevents the deposition of ash and also facilitates sulphation process to occur [52], which in turn reduces the corrosion impact [47]. In addition, the prism results in a uniform distribution of flue gas speed, temperature and oxygen, thereby preventing the creation of hot spots within the boiler [32]. Moreover, the prism removes heat from the combustion zone, and reduces the temperature, which then reduces the volatilization of alkali chlorides, leading to generation of less chemically corrosive deposits [47, 52].

6. Future perspectives and knowledge gaps

The present chapter aims for an investigation, understanding, and analysis of the factors that lead to unsustainable treatment of MSW in incinerators. Several research directions have been identified that may help in the improvement of waste treatment in the waste to energy incinerators.

High temperature corrosion is a problematic issue that is hindering the sustainable and efficient treatment of waste in incinerators. The general theory of the corrosion process is the active oxidation. Most researchers studied the global corrosion reaction process. However, studies on the elementary reactions and their reaction rate constant which actually shows how the corrosion process occurs are scarce and this can be achieved via kinetic modeling of corrosion on various alloys as well as various operating conditions which needs to be validated by full-scale experimental investigations. Nonetheless, the operation conditions in full scale set up are difficult to control. In light of that, lab scale experiment can be used to validate the modeling. This may help plant operators to select the operating conditions that result in low corrosion rate as well as on alloy which can be used as superheater thereby improve the treatment efficiency of waste.

More research is needed on the optimization of the incineration conditions through the use of various technologies including Ecotube and Segher boiler prism which reduces corrosion and ash deposition by ensuring high turbulence and optimal mixing within the boiler but studies showing these technologies are rare.

7. Conclusions

Waste-to-energy incinerators have been generally accepted as an environmentally preferred method for waste management. However, sustainable and efficient treatment of waste in incinerators is hindered by high temperature corrosion and ash

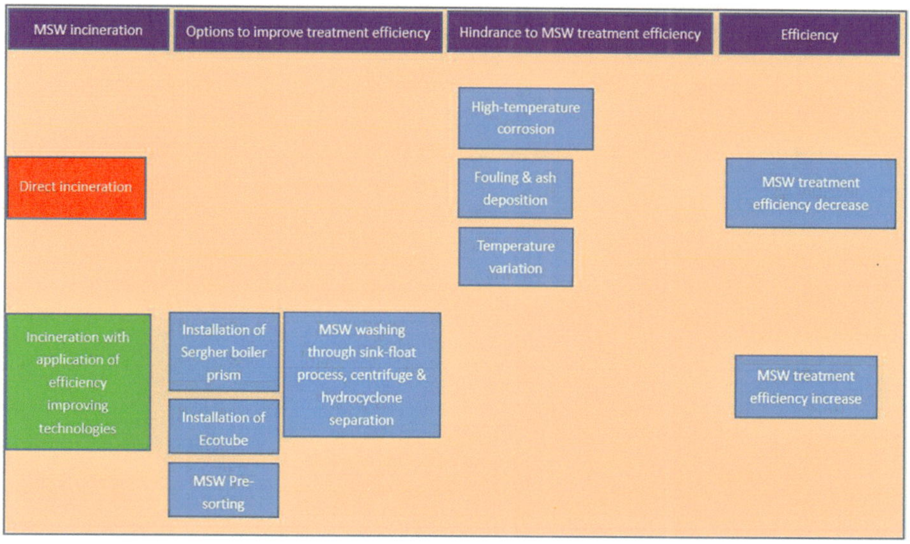

Figure 4.
Framework for options to improve MSW treatment efficiency.

deposition. This studies evaluated the potential for improving the efficient treatment of waste in incinerators. It is found that optimizing the operating conditions during incineration reduces the hindrance problem of high temperature corrosion and ash deposition. The operating conditions can be optimized by conducting a kinetic modeling which identifies the conditions that leads to reduced corrosion rate. Also, use of new technologies such the ecotube and sergher boiler prism which ensures high turbulence and optimal mixing within the boiler reduces the corrosion ad ash deposition problem, thereby improving the incineration of waste. Moreover, sorting of waste and related feedstocks improves the treatment of waste in incinerators as chlorine containing waste fractions may have been removed and this can be achieved through the use of technologies such as the max Al robotic sorter. Furthermore, to efficiently treat MSW in incinerators, chlorine load prior to waste incineration should be reduced by washing alkali chlorides and separating plastic from the waste as received. **Figure 4** summarizes the hindrances and options to improve MSW treatment efficiency during incineration.

Acknowledgements

The authors are thankful to the IntechOpen editor for the invitation to contribute this chapter of the book. Also, the author is thankful to Prof. Gwenzi affiliated at University of Zimbabwe for the insightful guidance during the writing of this chapter.

Conflict of interest

The author declare that there is no conflict of interest.

Author details

Terrence Wenga
Faculty of Agriculture Environment and Food Systems, Department of Soil Science and Environment, University of Zimbabwe, Harare, Zimbabwe

*Address all correspondence to: wengat@yahoo.com

References

[1] Ma W, Wenga T, Frandsen FJ, Yan B, Chen G. The fate of chlorine during MSW incineration: Vaporization, transformation, deposition, corrosion and remedies. Progress in Energy and Combustion Science. 2020;**76**:100789

[2] Varjani S, Shahbeig H, Popat K, Patel Z, Vyas S, Shah AV, et al. Sustainable management of municipal solid waste through waste-to-energy technologies. Bioresource Technology. 2022;**355**:127247

[3] Hoornweg D, Bhada-Tata P. What a Waste: A Global Review of Solid Waste Management. 2012

[4] Chua HS, Bashir MJK, Tan KT, Chua HS. A sustainable pyrolysis technology for the treatment of municipal solid waste in Malaysia. In: AIP Conference Proceedings of 6TH International Conference on Environment (ICENV2018); 11-13 December 2018; Penang, Malaysia. Vol. 2124, No. 1. United States of America: AIP; 2019. p. 020016

[5] Wang Y, Zhang X, Liao W, Wu J, Yang X, Shui W, et al. Investigating impact of waste reuse on the sustainability of municipal solid waste (MSW) incineration industry using emergy approach: A case study from Sichuan province, China. Waste Management. 2018;**77**:252-267

[6] Navigant-Research. Global Waste-to-Energy Market to Reach $29.2 Billion by 2022. Available from: https://www.navigantresearch.com/newsroom/global-waste-to-energy-market-to-reach-29-2-billion-by-2022 [Accessed: February 2022]

[7] IRENA. Renewable Energy Statistics 2017 [Internet]. 2022. Available from: https://www.irena.org/-/media/Files/IRENA/Agency/Publication/2017/Jul/IRENA_Renewable_Energy_Statistics_2017.pdf [Accessed: October 20, 2022]

[8] Pedersen AJ, van Lith SC, Frandsen FJ, Steinsen SD, Holgersen LB. Release to the gas phase of metals, S and Cl during combustion of dedicated waste fractions. Fuel Processing Technology. 2010;**91**(9):1062-1072

[9] Bogner J, Ahmed MA, Diaz C, Faaij A, Gao Q, Hashimoto S, et al. Waste Management. In: Metz B et al., editors. Climate Change 2007: Mitigation. Contribution of Working Group III to the Fourth Assessment Report of the Intergovernmental Panel on Climate Change. Cambridge: United Kingdom and New York, NY, USA: Cambridge University Press; 2007. pp. 587-613

[10] Coda B, Aho M, Berger R, Hein KRG. Behavior of chlorine and enrichment of risky elements in bubbling fluidized bed combustion of biomass and waste assisted by additives. Energy & Fuels. 2001;**15**(3):680-690

[11] CEWEP. Energising Waste-a Win-Win Situation [Internet]. 2022. Available from: https://www.cewep.eu/wp-content/uploads/2017/09/Energy-win-win-paper-April-2020.pdf [Accessed: October 20, 2022]

[12] Kumar A, Samadder SR. A review on technological options of waste to energy for effective management of municipal solid waste. Waste Management. 2017;**69**:407-422

[13] Prajapati P, Varjani S, Singhania RR, Patel AK, Awasthi MK, Sindhu R, et al. Critical review on technological advancements for effective waste

management of municipal solid waste—Updates and way forward. Environmental Technology & Innovation. 2021;**23**:101749

[14] Ionescu G, Rada E. Material and energy recovery in a municipal solid waste system: Practical applicability. International Journal of Environment and Resource. 2012;**1**(1):26-30

[15] Shah AV, Srivastava VK, Mohanty SS, Varjani S. Municipal solid waste as a sustainable resource for energy production: State-of-the-art review. Journal of Environmental Chemical Engineering. 2021;**9**(4):105717

[16] Abbasi SA. The myth and the reality of energy recovery from municipal solid waste. Energy, Sustainability and Society. 2018;**8**(1):36

[17] USDOE. Waste-to-Energy from Municipal Solid Waste. Office of Energy Efficiency and Renewable [Internet]. 2022. Available from: https://www.energy.gov/sites/prod/files/2019/08/f66/BETO--Waste-to-Energy-Report-August--2019.pdf [Accessed: October 20, 2022]

[18] CIWM. The R1 Energy Efficiency Formula [Internet]. 2022. Available from: https://www.ciwm.co.uk/ciwm/knowledge/the-r1-energy-efficiency-formula.aspx [Accessed: October 20, 2022]

[19] Chen D, Christensen TH. Life-cycle assessment (EASEWASTE) of two municipal solid waste incineration technologies in China. Waste Management & Research. 2010;**28**(6):508-519

[20] Hossain MS, Haque MA, Hoyos LR. Dynamic properties of municipal solid waste in bioreactor landfills with degradation. Geotechnical and Geological Engineering. 2010;**28**(4):391-403

[21] Barker K. Global municipal solid waste continues to grow [Internet]. 2022. Available from: https://www.recyclingproductnews.com/article/2395/global-municipal-solid-waste-continues-to-grow [Accessed: October 20, 2022]

[22] EPA. Energy Recovery from the Combustion of Municipal Solid Waste (MSW) [Internet]. 2022. Available from: https://www.epa.gov/smm/energy-recovery-combustion-municipal-solid-waste-msw [Accessed: October 20, 2022]

[23] Fruergaard T, Christensen TH, Astrup T. Energy recovery from waste incineration: Assessing the importance of district heating networks. Waste Management. 2010;**30**:1264-1272

[24] Xu H, Lin W, Romagnoli A. Technological review on enhancing the energy efficiency of MSW incineration plant. In: Proceeding of the IEEE Asian Conference on Energy, Power and Transportation Electrification (ACEPT); 30 October-2 November 2018; Singapore, New York: IEEE; 2018. pp. 1-7

[25] Porteous A. Why energy from waste incineration is an essential component of environmentally responsible waste management. Waste Management. 2005;**25**(4):451-459

[26] Grabke HJ. Fundamental mechanisms of the attack of chlorine, HCl on steels and high temperature alloys in the temperature range 400 °C and 900 °C. In: International Conference on Fireside Problems while Incinerating Municipal and Industrial Waste. Florida, USA: The Sheraton Palm Coast; 1989

[27] Redmakers P, Hesseling W, van de Wetering J. Review on corrosion in waste incinerators and possible effect of bromine. TNO report I02/01333/RAD. 2002. pp. 1-51

[28] Grabke HJ, Reese E, Spiegel M. The effects of chlorides, hydrogen chloride, and sulfur dioxide in the oxidation of steels below deposits. Corrosion Science. 1995;**37**(7):1023-1043

[29] Uusitalo MA, Vuoristo PMJ, Mäntylä TA. High temperature corrosion of coatings and boiler steels below chlorine-containing salt deposits. Corrosion Science. 2004;**46**(6):1311-1331

[30] Zahs A, Spiegel M, Grabke HJ. Chloridation and oxidation of iron, chromium, nickel and their alloys in chloridizing and oxidizing atmospheres at 400-700°C. Corrosion Science. 2000;**42**(6):1093-1122

[31] Albina DO. Theory and experience on corrosion of Waterwall and Superheater tubes of waste-to-energy facilities [MSC thesis]. Department of Earth and Environmental Engineering. Columbia University; 2005

[32] Lee SH, Themelis NJ, Castaldi MJ. High-temperature corrosion in waste-to-energy boilers. Journal of Thermal Spray Technology. 2007;**16**(1):104-110

[33] Li YS, Spiegel M, Shimada S. Corrosion behaviour of various model alloys with NaCl–KCl coating. Materials Chemistry & Physics. 2005;**93**(1):217-223

[34] Glarborg P, Marshall P. Mechanism and modeling of the formation of gaseous alkali sulfates. Combustion and Flame. 2005;**141**(1-2):22-39

[35] Otsuka N. A thermodynamic approach on vapor-condensation of corrosive salts from flue gas on boiler tubes in waste incinerators. Corrosion Science. 2008;**50**(6):1627-1636

[36] Albina DO, Millrath K, Themelis NJ. Effects of feed composition on boiler corrosion in waste-to-energy plants. In: Proceedings of NAWTEC12 12th Annual North American Waste-to-Energy Conference; 17-19 May 2004, Savannah, Georgia, USA: ASME; 2004. pp. 99-109

[37] Becidan M, Sørum L, Frandsen F, Pedersen AJ. Corrosion in waste-fired boilers: A thermodynamic study. Fuel. 2009;**88**(4):595-604

[38] Rammer B, Galetz MC. Kinetics of volatilization of high temperature corrosion products and its application to chlorine corrosion. Materials and Corrosion. 2017;**68**(2):186-196

[39] Soustelle M. An Introduction to Chemical Kinetics. Hoboken; USA: John Wiley and Sons, Inc.; 2013. pp. 250-290

[40] Ma W, Wenga T, Chen G. Kinetic modeling and experimental validation on the effect of KCl and SO2 concentration on corrosion of pure Fe under simulated municipal solid waste combustion. Energy Procedia. 2018;**152**:1302-1309

[41] Chen G, Wenga T, Ma W, Lin F. Theoretical and experimental study of gas-phase corrosion attack of Fe under simulated municipal solid waste combustion: Influence of KCl, SO2, HCl, and H2O vapour. Applied Energy. 2019;**247**:630-642

[42] Miller PD, Krause HH, Vaughan DA, Boyd WK. The mechanism in high temperature corrosion in municipal incinerators. Corrosion. 1972;**28**(7):274-282

[43] Oh H, Annamalai K, Sweeten JM. Effects of ash fouling on heat transfer during combustion of cattle biomass in a small-scale boiler burner facility under unsteady transition conditions. International Journal of Energy Research. 2011;**35**(14):1236-1249

[44] Baxter LL. Ash deposition during biomass and coal combustion: A

mechanistic approach. Biomass and Bioenergy. 1993;**4**(2):85-102

[45] Hansen LA, Nielsen HP, Frandsen FJ, Dam-Johansen K, Hørlyck S, Karlsson A. Influence of deposit formation on corrosion at a straw-fired boiler. Fuel Processing Technology. 2000;**64**(1):189-209

[46] Li Q, Zhou H. Effects of ash deposition and slagging on heat transfer. In: Zhang Y, editor. Theory and Calculation of Heat Transfer in Furnaces. Elsevier; 2016. pp. 173-191

[47] Adams B, Diederen HSW, Peeters K, Wijnhoven JPF, Eeraerts D. Seghers Boiler Prism: A Proven Primary Measure against High Temperature Boiler Corrosion. in North American Waste-to-Energy Conference. 2004

[48] Delay I, Swithenbank J, Argent BB. Prediction of the distribution of alkali and trace elements between the condensed and gaseous phases generated during clinical waste incineration. Journal of Alloys & Compounds. 2001;**320**(2):282-295

[49] Jensen PA, Sander B, Dam-Johansen K. Removal of K and Cl by leaching of straw char. Biomass & Bioenergy. 2001;**20**(6):447-457

[50] Chen HL, Pagano M. The removal of chlorine from Illinois coal by high temperature leaching. Fuel Processing Technology. 1986;**13**(3):261-269

[51] Hwang IH, Matsuto T, Tanaka N. Water-soluble characteristics of chlorine in char derived from municipal solid waste. Waste Management. 2006;**26**:571-579

[52] Grillo LM. Municipal Solid Waste (MSW) Combustion Plants. 2013. pp. 72-97

Chapter 6

Estimating the Calorific Value and Potential of Electrical Energy Recovery of Organic Fraction of Municipal Solid Waste through Empirically Equations and Theoretically Way

Gunamantha Made

Abstract

This chapter aimed to estimate the calorific value and potential of electrical energy that can be generated from the organic fraction of municipal solid waste through mathematical models, bioconversion, and thermochemical approach. The calorific values were calculated using the empirical relationship between higher heating value and ultimate analysis data, stoichiometric manner, and thermochemistry concept. The potential of electrical energy that can be produced was calculated based on literature data on the specified power plant. The result showed that the calorific value and potential of electrical energy recovery of the thermochemical approach are higher than others.

Keywords: municipal solid waste, organic fraction, higher heating value, energy, electrical energy recovery

1. Introduction

The increase in population in urban areas results in an increase in the amount of municipal solid waste (MSW) generated. AusAID [1] reported that a total of 38.5 million tons of solid waste were generated annually by the 232 million inhabitants in Indonesia (450 gm per person per day), of the which, 21.2 million tons contributed by the inhabitants of the island of Java. The 26 biggest Cities in Indonesia inhabit a totally 40.1 million people, generating in total an estimated 14.1 million tons per year (about 1 kg per person per day). In this country, municipal waste is composed of 62% of mainly organic waste, 14% plastics, 9% paper, 2% glass, 2% rubber and leather, 2% metals, and 13% of other waste types [1].

These physical characteristics of solid waste are indicated as a potential source of biomass mainly their bioorganic contents. Biomass can be considered as the solar energy stored in chemical bonds of organic material [2]. If the bond between carbon and hydrogen, and oxygen is broken down through decomposition, combustion, or decomposition process for these materials, the chemical energy stored or potential energy will be released [2]. Therefore, its energy content will be influenced by the elementary composition.

To determine the potential energy stored in the biomass, traditionally by direct measurement or theoretical approach. Direct measurements can be carried out by an experiment using a bomb calorimeter [3] and the theoretical can be determined from their elementary content such as carbon, hydrogen, oxygen, nitrogen, and sulfur [4]. In this, to analyze the potential of energy recovery required adequate biomass characteristics, especially in elemental compositions [4, 5]. In relation to the utilization of solid waste as an energy source, the investigation of their chemical elemental characteristics is beneficial to the suitable choice of energy conversion technologies including bioconversion, incineration, or thermochemical conversion processes.

However, there is lack of data associated with the chemical elemental characteristics of solid waste in Indonesia. Based on their elemental chemical characteristics, the energy content of solid waste fuel can be calculated from the heat of combustion. In the combusting, fuel will release its energy potential. The energy released or the heat of fuel combustion is heat when a fuel undergoes complete combustion with oxygen under standard conditions. It can be calculated as the difference between the heat of the formation of the products and reactants. In this, a chemical equation is required. The heat of the formation of the products and reactants can be obtained through a thermochemical table [4]. This change of the enthalpy approach is a thermochemical conversion process based. Estimation of the potential energy from organic waste fuel can also be determined by using the Buswell equation. Buswell [6] suggested a general equation for the anaerobic bioconversion process of biodegradable organic matter. This second way either requires stoichiometric equations for the bioconversion process or the biodegradability of waste. The biodegradability of organic solid waste can be expected from its volatile solid and lignin content [6].

Besides based on the theoretical way, the elementary characteristics data can also be used to estimate the calorific value or energy content based on their empirical relationship. Researchers in several countries have carried out extensive research to determine the empirical relation between the elemental and calorific value of biomass and solid waste fuel [4, 7–12]. The energy content of the estimation results can then be used to estimate the quantity of energy that can be recovered on fuel. Therefore, both through direct measurement, stoichiometric approach, as well as the empirical relationship can be used as a basis for estimating energy recovery from organic solid waste.

This paper aims to estimate energy content and recovery potential organic component of MSW by using experiment, theoretical, and empirical approaches. The energy recovery potential is determined by the combustion and anaerobic digestion process.

2. Methodology

There were four steps used. First, ultimate analysis data and calorific value of the organic fraction of MSW (OFMSW) were collected based on literature data [12, 13]. Second, the hypothetical chemical formula was calculated from their ultimate analysis data as a basis for estimating the energy released when thermal conversion and bioconversion processes were applied. Third, correlations between higher heating value (HHV) and ultimate analysis data were developed. Fourth, comparing the potential energy recovered from experimental, stoichiometric manner, and empirical models developed.

2.1 Characteristics of waste

The characteristics of OFMSW data such as ultimate analysis included carbon (C), hydrogen (H), nitrogen (N), oxygen (O), sulfur (S) (mass percentages on a dry basis), and water content (H2O, mass percentages as discarded), calorific value, volatile solid (VS), and lignin content were used. As reported in [12], the organic fraction of MSW was collected from a landfill site in Indonesia. The content of carbon (C) was determined by the standard American Standard for Testing Materials ASTM) D 5373, hydrogen (H) by the standard ASTM D 5373, nitrogen (N) by the standard ASTM D 5373, oxygen (O) by standard ASTM D 3176, and sulfur (S) by standard ASTM D 4239. These elemental analysis data were performed using CHNS-O Analyzers (Perkin-Elmer 2400 Series) at the mineral and coal technology laboratory, Bandung Indonesia [12]. The higher heating value (HHV) of the sample was determined based on the standard ASTM D 5865 by using Bomb Calorimeter [12]. The lignin content was analyzed at Food and Nutrient Laboratory Gadjah Mada of University Yogyakarta [13].

2.2 Hypothetical development chemical formula

The molecular formula is the actual whole number ratio between the elements. The quantity of each element was divided by its molar mass to give the number of hypothetical moles of each element. Furthermore, these mole ratios were used to determine the molecular formula. The molecular formula was used too in determining the principal reactions that occur during combustion and bioconversion.

2.3 Empirical correlations

Correlations between calorific value with percent carbon (C), hydrogen (H), oxygen (O), nitrogen (N), and sulfur (S) were used to estimate calorific value. In this case, correlations equations were developed from previously available empirical equations (**Table 1**).

The fulfillment of model performances was assessed by four statistical criteria, namely: sample paired test, coefficient of correlation, average absolute error (AAE), and average bias error (ABE). Sample paired test and coefficient of correlation were determined by SPSS 17 version. The symmetrical relationships were used to determine the correlation between the calorific value of the calculation results of the model with the measurement. The coefficient correlations were used to determine strong or weak relationships by the following criteria: 0.00–0.199 very weak, 0.20–0.399 weak, 0.40–0.599 sufficient, 0.60–0.799 strong, and very strong 0.80–1.00.

The average absolute error can be expressed as follow:

$$AAE = \frac{1}{n}\sum_{i=1}^{n}\left|\frac{Ya - Yp}{Ya}\right| \times 100\% \tag{11}$$

Where AAE is the mean percentage of absolute error, Ya and Yp are the actual and expected values. n represents the number of data points. The sum of squared errors can be given by the following equation:

Eq.	Models	Sample	Ref.
(1)	$HHV\ (kJ/kg) = -1.494 + 0.474C - 0.803H + 0.034O + 0.982N$	OFMSW	[12]
(2)	$HHV\ (kJ/kg) = -1.309 + 0.475C - 0.796H + 0.031O + 1.008N - 1.395S$	OFMSW	[12]
(3)	$HHV\ (kJ/kg) = -1.3675 + 0.3137C + 0 : 7009H + 0.0318O$	Biomass	[10]
(4)	$HHV\ (kJ/kg) = 0.399C + 1.4H - 0.139O + 0{,}105S$	MSW	[14]
(5)	$HHV\ (kcal/kg) = 81C + 342.5(H - 1/8O) + 22.5S$	MSW and RDF	[7]
(6)	$HHV\ (kcal/kg) = 81(C - 3/8O) + 171/8O + 342.5(H - 1/16O) + 25S$	MSW and RDF	[7]
(7)	$HHV\ (kcal/kg) = 81(C - 3/8O) + 342.5H + 22.55S + 171/4O$	MSW and RDF	[7]
(8)	$HHV\ (Mj/kg) = 0.327C + 1.241H - 0.089O - 0.26N + 0.074S$	MSW	[4]
(9)	$HHV\ (kJ/kg) = (35.160C + 116.225H - 11.090O + 6.280N + 10.465S) \times 10^{-2}$	Fossil fuel	[15]
(10)	$HHV\ (kJ/kg) = (340.39\ C + 1320.83\ H + 68.30\ S - 15.28\ ash - 118.5\ (O + N)) \times 10^{-3}$	Coal	[16]

Note: RDF = refuse-derived fuel.

Table 1.
Correlation models from any research.

$$ABE = \frac{1}{n}\sum_{i=1}^{n}\frac{Ya - Yp}{Ya} \times 100\% \tag{12}$$

Where ABE is the mean percentage of bias error.

3. Results and discussion

3.1 Data analysis ultimate and calorific value

Data ultimate analysis of the organic fraction of MSW that is used in this chapter can be seen in **Table 2**. The database showed a variation content of C, H, N, and S. The waste contained 38.95% carbon. This value has a good agreement with the values reported in the literature. The average percentage of hydrogen is 5.44%. Compared with other studies, the value of hydrogen is higher than has been reported in [9, 17]. **Table 2** also listed that the average percentage of available nitrogen is nearly 1.47%. The content of nitrogen was reported higher than the value reported by [9, 10, 17] and the average percentage of sulfur is 0.12%. This value is relatively low compared to the results presented in [6].

Table 2 also presented the calorific value of waste. The calorific value (CV) of a material indicates the energy content or the heat released when it is burnt in the presence of air. CV can be measured as energy content per unit mass or volume; kJ/kg for solids, MJ/L for liquid, or MJ/Nm^3 for gas fuel. CV of fuel can be expressed in two forms: gross CV (GCV), or a higher heating value (HHV), and net CV (NCV), or lower heating value (LHV) [10]. HHV is the total energy content that is released when a fuel is burned in air, including the latent heat contained in the water vapor, and therefore represents the maximum amount of potential energy that can be charged from a source of fuel. The actual amount of energy that can be collected will vary depending on the form of fuel and conversion technologies used [6]. In this, the calorific value in **Table 2** is the HHV of waste. The average HHV was found 15.41 MJ/kg with a standard deviation of 1.98. It also observed that the calorific value is lower than the calorific value of the lignocellulosic as well as cellulosic biomass. In [18], it was reported that the cellulosic biomass has an average calorific value of 17.73 MJ/kg and for lignocellulosic materials 26.7 MJ/kg.

3.2 Hypothetical chemical formula for organic component of MSW

The chemical formula of OFMSW can be approximated as a hypothetical compound of the form $C_aH_bO_cN_dS_e$ [6, 19–24]. **Table 3** shows the determination of the hypothetical chemical formula of OFMSW using the average value of ultimate data. By using S as a base, then, the empirical chemical formula is $C_{842}H_{1411}O_{641}N_{27}S$. However, in view of sulfur and nitrogen is relatively small components of it, if the nitrogen and sulfur are removed, the molecular structure of the waste is very close to the cellulose ($C_6H_{10}O_5$). On the other hand, in [22], it was reported that the structure of mixed food and plant wastes can be approximated by the molecular composition ($C_6H_{10}O_4$). The molecular compositions are useful to determine fundamentals reaction during thermal conversion or bioconversion process. Therefore, if the structure varies significantly then the range enthalpy of combustion will also vary, thus, increasing the uncertainty associated with the quantity of energy that can be

Sample	Elements composition (%)					Ash (%)	Calorific value (MJ/kg)
	C	H	N	S	O		
1	39.37	5.71	1.87	0.18	37.63	15.24	15.42
2	42.31	5.98	1.69	0.13	38.40	11.49	16.64
3	24.63	3.96	1.04	0.13	29.56	40.68	8.91
4	25.81	4.48	1.11	0.13	34.82	33.65	9.84
5	39.65	5.70	1.87	0.19	39.62	12.97	15.8
6	42.89	6.06	1.60	0.11	39.27	9.97	17.00
7	39.99	6.07	1.69	0.10	39.89	12.26	15.77
8	40.27	6.04	1.71	0.09	39.44	12.45	15.82
9	39.22	5.86	1.75	0.11	39.50	13.56	15.43
10	40.15	5.99	1.80	0.14	40.06	11.86	15.92
11	40.34	6.07	1.74	0.11	40.59	11.15	15.90
12	38.94	5.79	1.73	0.14	30.43	14.97	15.19
13	41.47	5.74	0.74	0.11	42.48	9.46	15.76
14	40.14	5.59	1.02	0.15	40.99	12.11	15.26
15	38.68	5.38	0.74	0.02	40.39	14.79	14.53
16	39.89	5.52	1.12	0.17	40.12	13.18	15.27
17	40.57	5.68	0.74	0.11	41.71	11.19	15.58
18	40.29	5.58	1.08	0.17	41.16	11.72	15.56
19	40.11	4.85	1.83	0.12	42.43	10.66	16.80
20	41.16	4.93	1.68	0.09	42.33	9.81	16.99
21	39.45	4.86	1.66	0.11	41.75	12.17	16.60
22	39.66	4.87	1.65	0.13	41.43	12.26	16.52
23	39.19	4.86	1.7	0.11	41.97	12.17	16.48
24	40.51	4.91	1.78	0.11	42.26	10.43	16.92
Average	38.95	5.44	1.47	0.12	39.51	14.18	15.41
Deviation Standard	4.34	0.58	0.40	0.04	3.42	7.32	1.98

Table 2.
Ultimate and calorific value organic fraction of MSW (dry basis).

recovered from the waste stream. Conversely, if the structure appears to be fairly stable, i.e. not deviating from the mean greatly, then the enthalpy of formation is likely to be fairly constant [24]. It supported the claim that the compound $C_6H_{10}O_5$ can be used to approximate the chemical structure of OFMSW in Bali. The standard enthalpies of formation and combustion can be used to deduce an approximate heating value of $C_6H_{10}O_5$, as well as the heating value, can be estimated from the biogas generated when the anaerobic decomposition was applied.

The estimated heat of the combustion reaction can be performed using the thermochemical table. The use of thermochemical data tables from textbooks that are generally located on the back page is a general way. For example, the large amounts of

Items	Elemental data (dry basis)				
	Carbon (C)	Hydrogen (H)	Oxygen (O)	Nitrogen (N)	Sulfur (S)
Composition (%)	38.95	5.44	39.51	1.47	0.12
Atomic weight (kg/kmol)	12.00	1.00	16.00	14.00	32.00
mol	3.25	5.44	2.47	0.11	0.00
mol ratio (S basis)	842.06	1410.60	640.70	27.29	1
mol ratio (C basis)	6	10.05	4.57	0.20	0.01

Table 3.
Determined hypothetical chemical formula organic fraction of MSW.

standard heat of formation data can be found in Perry's Chemical Engineering Handbook or Handbook of Chemistry and Physics. Based on these data, the enthalpy change of the reaction can be predicted in various chemical reactions. The standard heat of the formation of a compound is the enthalpy change associated with the formation of 1 mol of a compound from its elements in their standard state at temperatures of 25°C and a pressure of 1 bar. The standard enthalpy of formation for H_2O (l) and CO_2 (g) is successively −285.830 kJ/mol and − 393.509 kJ/mol, while cellulose 733 kJ/mol. The standard enthalpy change for the combustion of the hypothetical chemical formula can be determined by finding the difference between the standard enthalpy of the formation of products and reactants. In the case of cellulose, the difference is.

$$\Delta H_{ocomb} = 5(-393.509) + 6(-285.830) - (-733) = -2949.5\ \text{kJ/mol} = -18.21\ \text{MJ/kg}.$$

Estimating the heating value through the bioconversion process can be performed using the Buswell equation. Buswell in 1952 found an equation for estimating the products of the anaerobic decomposition of organic materials with the general chemical composition $C_cH_hO_oN_nS_s$. By following the pattern of the Buswell equation, the equation anaerobic decomposition of hypothetical predetermined formula can be expressed by Eq. (13) below:

$$C_{842}H_{1411}O_{641}N_{27}S + 188.5\ H_2O - - - -412.25\ CO_2 + 426.75\ CH_4 + 27\ NH_3 + H_2S \tag{13}$$

Eq. (13) shows that, 1 (one) ton or 0.05 kg mol organic fraction can produce 19.239 kg mol or 431 m^3 CH_4 (STP). With regard to the energy content of methane is 36 MJ/m^3, then, 1 ton (1000 kg) organic fraction of MSW has the potential of energy recovery 15.516 MJ/kg. These results assumed the substrate is perfectly biodegradable. The result is also not much different from the average calorific value obtained from the measurement results (15.41 MJ/kg) as mentioned in the previous section. With reference to the data reported by [13], the average biodegradability of the organic waste component in their study area was 0.16. Based on the data biodegradability, hence, net CH_4 generated from the equation Buswell is 68.96 m^3. Thus, 1 ton (1000 kg) organic fraction of MSW has the potential of energy recovery only 2.482 MJ/kg.

Another possible way to estimate methane production is by using the carbon content of the organic fraction of MSW as shown in Eq. (14). Carbon content (C) expressed with gC. BF is the biodegradability of organic waste component, 0.5 is the value set for the volumetric ratio of methane to the total landfill gas, 22.4/12 is the conversion factor C into a gas at 1 atm and 25°C, expressed as L CH_4/gC. Lo expressed as m^3 CH_4 [13].

$$Lo = C \times BF \times 0.5 \times 22.4/12 \times 1000 \quad (14)$$

3.3 Predicting HHV from ultimate analysis data

Several empirical correlations have been developed to estimate the HHV of various biomass and fossil fuels using ultimate analysis data. The correlations were generally derived from the equation for coal fuel developed by Dulong [9]. According to the concept that the fuel is essentially organic matter that has potential energy because of the carbon-hydrogen and carbon–oxygen bonds, the HHV relationship with the ultimate analysis data can also be adopted on other fuels, including solid waste HHV is a function of carbon, hydrogen, oxygen, and nitrogen have been developed for fuel garbage [6]. The development of empirical models based on multiple linear regression using the content (in% dry weight) of carbon, hydrogen, oxygen, nitrogen, and sulfur as the independent variable and the Higher Heating Value (HHV) as the dependent variable will help in determining the contribution each element in the ultimate analysis data to predict the HHV of waste. The assessment of developed models by several researchers (**Table 1**) was conducted in this section based on the results calculation of developed models obtained from the experiment. The results summary is given in **Table 4**.

According to the t-test value with 5% significance level, **Table 3** shows that there is no significant difference in calorific value between the model's calculation with obtained from the experiment because the value of the t-test is within the range of t-table −2.069 to 2.069. This implies that there were no measurement errors in bomb calorimeter operation in this ultimate analysis. Based on the correlation coefficient, only two equations (Eqs. (5) and (6) do not show a very strong correlation between the results of calculations with the experiment. Otherwise, based on the average value of absolute and bias error, only Eq. (1) and (2) are both less than 5%. However, in determining the energy that can be collected from the organic fraction of MSW, the estimation of Eq. (1), (2), (8), (9), and (10) are used due to their errors less than 10%.

3.4 Estimation of energy recovery

Based on estimation calorific values obtained from empirical relationships elected, thermochemical and biochemical processes through stoichiometric equations, carbon content, and direct measurement, the potential electrical energy recovery was determined. The results are summarized in **Table 5**. The value is obtained by assuming that the heat generated from the flue gases of combustion can be utilized to generate steam from the boiler. In this context, the boiler is an integral part of the conversion system. However, the efficiency of boiler performance varies according to the energy source. According to [6], the variables that determine the efficiency of the boiler included the energy content of the fuel, moisture content, flue gas temperature, and the inner physical design of the boiler. In this chapter, the efficiency boiler is set to 70% by adopting the value in [6] for mass burning. Similarly, from the same literature, the efficiency factor of steam turbines and electrical generators data were also used to estimate the electric power generated. Besides that, the facility consumption and loss using the data available in [6]. In the system of bioconversion, the production of

DOI: http://dx.doi.org/10.5772/intechopen.108232

Eq.	Models	Correlation	t-test	Error	
				AAE	ABE
(1)	HHV (kJ/kg) $= -1.494 + 0.474C - 0.803H + 0.034O + 0.982N$	0.997	0.501	0.932	0.182
(2)	HHV (kJ/kg) $= -1.309 + 0.475C - 0.796H + 0.031O + 1.008N - 1.395S$	0.997	0.689	0.890	0.125
(3)	HHV (kJ/kg) $= -1.3675 + 0.3137C + 0 : 7009H + 0.0318O$	0.919	0.004	6.052	−3.728
(4)	HHV (kJ/kg) $= 0.399C + 1.4H - 0.139\ O + 0{,}105S$	0.816	0.000	15.172	−15.172
(5)	HHV (kcal/kg) $= 81C + 342.5(H - 1/8\ O) + 22.5S$	0.748	0.000	9.876	9.406
(6)	HHV (kcal/kg) $= 81(C - 3/8O) + 171/8O + 342.5(H - 1/16O) + 25S$	0.791	0.045	8.623	−4.055
(7)	HHV (kcal/kg) $= 81(C - 3/8O) + 342.5H + 22.55S + 171/4O$	0.881	0.000	50.438	−50.438
(8)	HHV (MJ/kg) $= 0.327C + 1.241H - 0.089O - 0.26N + 0074S$	0.813	0.460	6.949	−1.695
(9)	HHV (kJ/kg) $= (35.160\ C + 116.225\ H - 11.090\ O + 6.280\ N + 10.465\ S) \times 10^{-2}$	0.831	0.171	7.103	−2.558
(10)	HHV (kJ/kg) $= (340.39\ C + 1320.83\ H + 68.30\ S - 15.28\ ash - 118.5\ (O + N)) \times 10^{-3}$	0.807	0.683	6.742	−1.093

Table 4.
Assessment empirical relationship between HHV and the ultimate analysis data.

Approach	Potential of energy (kJ/kg)	Parameters			Gross electrical energy generated (kWh/kg)*
		Boiler efficiency (%)	Turbine efficiency (%)	Generator efficiency (%)	
Empirical model (1)	15390	70	29	90	0.78
Empirical model (2)	15400	70	29	90	0.78
Empirical model (8)	15592	70	29	90	0.79
Empirical model (9)	15736	70	29	90	0.80
Empirical model (10)	15515	70	29	90	0.79
Thermochemical	18206	70	29	90	0.92
Bioconversion (Buswell)	2482	—	24	90	0.15
Carbon content (IPCC)	2199		24	90	0.13
Bomb calorimeter	15410	70	29	90	0.78

**1 kWh = 3600 kJ*

Table 5.
Estimation of energy recovery through thermal conversion and bioconversion.

methane gas from the process was considered to drive the gas turbine. The efficiency gas turbine (regenerative type) and the power generator was assumed 24% and 90% [6].

Table 5 expressed, the thermochemical approach giving the highest value. In this case, the approach was assumed as a complete combustion process. In addition, the calculation in this approach was based on the assumption that the molecular formula of waste is cellulose or without the involvement of S and N. Thus, based on the electrical energy that can be generated, the thermal conversion process tends to be more advantageous than the bioconversion process. Energy potential generated from the thermal conversion process reaches more than five times the bioconversion process. This is possible because the biodegradability of waste is too small. However, it is important to note that the calculation of the energy obtained through measurement using a bomb calorimeter should be closest to the real conditions. The others were estimation approaches. The estimation closest to the measurement results is of equations developed from the same sample.

4. Conclusions

Increased attention on MSW to energy has resulted in the increasing need for the characteristics and potential of this material. In Indonesia, this attention is more focused on the organic fraction of MSW or their biomass components. Various methods can be used to determine the energy potential and the amount of electrical energy that can be generated from the organic fraction of MSW. These methods can be developed through an empirical relationship between elementary constituents of waste by energy content, theoretical approach, or through measurement using a bomb calorimeter. This chapter concluded that each approach provides different results of energy recovery potential. The thermochemical process has given the highest calorific value (18.21 MJ/kg) and electric energy recovery (0.92 kWh/kg).

Author details

Gunamantha Made
Ganesha University of Education, Singaraja, Indonesia

*Address all correspondence to: md_gunamantha@yahoo.com

References

[1] DHV BV. Scoping study for solid waste management in Indonesia Technical Report. Indonesian Infrastructure Initiative: Australia Aid; 2011. www.indii.co.id

[2] Mckendry P. Energy production from biomass (part 1): Overview of biomass. Bioresource Technology. 2002;**83**:37-46

[3] Omalski ESD, Churney KL, Ledford AE, Colbert JC, Bruce SS, Buckley TJ, et al. Assessing the credibility of the calorific content of municipal solid waste. Pure and Applied Chemistry. 1991;**63**(10):1415-1418

[4] Akkaya EA, Demir A. Energy content estimation of municipal solid waste by multiple regression analysis. In: 5th International Advanced Technologies Symposium (IATS'09), May 13–15. Karabuk, Turkey; 2009

[5] Ruth AL. Energy from municipal solid waste: a comparison with coal combustion technology. Progress in Energy and Combustion Science. 1998;**24**:545-564

[6] Tchobanoglous G, Theisen H, Vigil S. Integrated Solid Waste Management, Mcgraw-Hill Series in Water Resources and Environmental Engineering. Mcgraw-Hill: New York; 1993

[7] Chang NB, Chang YH, Chen WC. Evaluation of heat value and its prediction for refused-derived fuel. Science of the Total Environment. 1997;**197**(1-3):139-148

[8] Meraz L, Oropeza M, Dominguez A. Prediction of the combustion enthalpy of municipal solid waste. The Chemical Educator. 2002;**7**:66-70

[9] Demirbas A. Combustion characteristics of different biomass fuels. Progess in Energy and Combustion Science. 2004;**30**:219-230

[10] Sheng C, Azevedo JLT. Estimating the higher heating value of biomass fuels from basic analysis data. Biomass and Bioenergy. 2005;**28**:499-507

[11] Telmo C, Lousada J, Moreira N. Proximate analysis, backwards stepwise regression between gross calorific value, ultimate and chemical analysis of wood. Bioresource Technology. 2010;**101**: 3808-3815

[12] Gunamantha M. Predicting the higher heating value of biogenic solid waste component from their ultimate analysis data. In: Proceeding(s) of National Seminar on Waste Management for Sustainable Development, Environmental Department, Sepuluh November Institute of Technology, February 21. Surabaya, Indonesia; 2012

[13] Gunamantha M, Yuningrat W. Inventory and greenhouse gas reduction potential of landfill in Bengkala Singaraja, Bali Province. International Journal of Environmental Science and Engineering Research (IJESER). 2012;**3**(3):170-179

[14] Ni-Bin Chang A, Davila E. Municipal solid waste characterizations and management strategies for the lower Rio Grande Valley. Texas, Waste Management. 2008;**28**:776-794

[15] Annamalai K, Sweeten M, Ramalingam SC. Estimation of gross heating value of biomass fuels. American Society of Agricultural Engineers. 1987; **30**(4):1205-1208

[16] Kök MV, Keskin C. Calorific value determination of coals by DTA and ASTM methods, comparative study. Journal of Thermal Analysis and Calorimetry. 2001;**64**:1265-1270

[17] Abu-Qudais M, Abu-Qdais HA. Energy content of municipal solid waste in Jordan and its potential utilization. Energy Conversion and Management. 2000;**41**:983-991

[18] Jenkins BM, Baxter LL, Miles TR Jr, Miles TR. Combustion properties of biomass. Fuel Processing Technology. 1998;**54**:17-46

[19] Tsiliyannis CA. Report: comparison of environmental impacts from solid waste treatment and disposal facilities. Waste Management & Research. 1999; **17**:231-241

[20] Vidal R, Gallardo A, Ferre J. Integrated analysis for pre-sorting and waste collection schemes implemented in Spanish cities. Waste Management & Research. 2001;**19**:380-391

[21] Themelis NJ, Hwan Kim Y, Brady MH. Energy recovery from New York City municipal solid wastes. Waste Management & Research. 2002;**20**: 223-233

[22] Themelis NJ, Kim YH. Material and energy balances in a large-scale aerobic bioconversion cell. Waste Management & Research. 2002;**20**:234-242

[23] Durmusoglu E, Corapcioglu MY, Tuncay K. Landfill settlement with decomposition and gas generation. Journal of Environmental Engineering. 2005;**131**:1311-1321

[24] Cester M, Stupples D, Lees M. A Comparison of the physical and chemical composition of UK waste streams based on hypothetical compound structure. In: 13th European Biosolids and Organic Resources Conference, 10–12 November. Manchester; 2008. http://Eprints.White rose.Ac.UK/8548/

Chapter 7

Anaerobic Digestion of Organic Solid Waste: Challenges Derived from Changes in the Feedstock

Ángeles Trujillo-Reyes, Sofía G. Cuéllar, David Jeison, Antonio Serrano, Soraya Zahedi and Fernando G. Fermoso

Abstract

Over the years, research on the anaerobic digestion of solid waste has mainly focused on single feedstocks with a fixed composition. Nevertheless, the impact assessment that drastic changes in the type and composition of feedstock might have on AD process stability has not been investigated in depth. The existence of a wide variety of organic solid waste whose generation and composition are highly dependent on seasonality, just as the possibility of using treatment plant facilities already in operation for treating new waste, makes it necessary to improve our knowledge of transitory states in AD. This chapter aims to provide insight into research on transitory states during the AD process when the type or composition of the feedstock has suffered a change to assess whether the AD process was finally able to adapt to system disturbances. Information about process stability control and microbial population adaptation, among others, derived from the transition states will be addressed.

Keywords: organic waste management, process stability, seasonality, substrate change, transient conditions

1. Introduction

High global population growth has led to an excessive increase in solid waste generation. According to the United Nations, at least 7000 and 10,000 million tons of solid waste are collected worldwide yearly [1]. The principal sectors responsible for this amount of solid waste are as follows: (1) construction and demolition (C&D) (34%); (2) municipal solid waste (MSW) (24%); (3) industrial (21%), and (4) commercial (11%) [1].

The MSW is the waste generated by households, mainly composed of an organic fraction, plastic, and paper. Although the organic fraction in MSW varies according to income levels countries (high-income countries ~20–40% and low-income countries ~50–70%), sociocultural patterns, and climatic factors, it could be considered that around half of the MSW would correspond to the organic fraction [2]. It means that about 840–1200 million tons of organic solid waste would be globally generated [1, 3]. It may even be argued that this amount of organic solid waste would be even higher as it does not consider the agro-industrial sector, whose waste is composed primarily of organic matter. The producers hardly report data from this sector. However, a rough estimate by the International Solid Waste Association (ISWA) concluded that approximately 10,000–20,000 million tons of waste are generated annually by crops, farms, vineyards, dairies, and other agri-food industries [1]. In summary, it could be stated that the amount of organic solid waste generated globally is highly significant, and its leadership in the waste generation field remains a relevant issue.

Globally, in 2018, about 19% of solid wastes were recycled and/or composted, 11% incinerated, 37% disposed of in landfills (with or without gas collection), and 33% disposed of in open dumps (uncontrolled waste disposal) [3]. The high percentage of waste derived from landfilling or open dumps indicates that there are still great opportunities for improvement in the management of organic solid waste. Within the technologies available for recycling and composting, anaerobic digestion (AD) has been recognized as an effective and interesting waste management technology, since it can produce green energy when converting organic matter from waste into biogas [4–6].

Increasing solid waste generation requires cost-effective and environmentally friendly processes, such as AD. With proper control of the AD process, this technique could be adapted to different operating conditions and changes in the feedstock. It would allow the possibility of treating a greater quantity and variability of seasonal waste (e.g. MSW; fruit and vegetable waste (FVW); and juice company waste) in a single solid waste plant, using existing facilities [7]. This advantage could even improve the biogas production and the economic viability of the plants [7, 8]. Because of these reasons, the ability of the AD process to adapt to different changes could be a compelling topic. In this context, variations of parameters have been extensively studied, such as organic loading rate (OLR), hydraulic retention time (HRT), or the operational temperature in the AD process [9–17].

Despite the existing knowledge about the previously cited operational parameters, feedstock type or composition changes in the AD process have been poorly investigated and could affect the process behavior. When an AD process is carried out under fixed operating conditions, there is a steady state in which the system conditions remain constant. However, when a change or disturbance affects the system, e.g. a change in the feedstock type or composition, the existing stability conditions can be lost, resulting in a transitory state. A transitory state could be defined as the period that elapses from when the change is applied to the system until system stabilization is reached. During this period, parameters such as the biogas production and composition or the concentration of volatile fatty acids (VFAs) could change because of biodegradability, pH, organic matter content, and other feedstocks' characteristics. This could directly affect the subsequent AD performance [7].

It has also been reported that the way how changes in conditions are made can cause different results in the system's adaptation [17]. For example, an aggressive change usually results in considerable instability in the system. On the contrary, a gradual change usually entails fewer fluctuations, because the system has more time to adapt to the new conditions. For this reason, the evaluation of the change influence on the development of the transitory state is a fundamental step to realize in the AD process [9].

This chapter aims to provide an insight into the available research on transitory states during the anaerobic digestion process when the feedstock type or composition has suffered a change, to assess the adaptation of the anaerobic digestion process to the system perturbations.

2. Monitorization of the anaerobic digestion process stability of organic solid waste during transitory states

This section describes the control and operational parameters relevant for determining the stability and the microbial population adaptation of AD processes. These factors have been reported as fundamental in the literature for transitory states during the AD processes, when the feedstock type or composition changes.

2.1 pH, alkalinity, and volatile fatty acids concentration

Monitoring parameters such as pH, alkalinity, and VFAs concentration are fundamental indicators of the equilibrium and stability of the AD process [6, 18].

Microbial growth and activity strongly depend on the environmental pH [19–21]. According to literature, methane producers are most active at a neutral pH, i.e. between 6.5 and 8.5 [22, 23], while at lower pH (5.0–6.0), its activity decreases severely, being active only for the acids-producing microorganisms [21]. If pH is rapidly increased or decreased concerning the existing environmental conditions, the microbial activities of specific microbial species could be inhibited. Notably, the methanogenic archaea inhibition would affect the activity of anaerobic microorganisms and, subsequently, the whole AD process performance [24]. The rapid increase or decrease of pH values could mostly occur for substrates from different origins whose physicochemical characteristics are not similar [4]. Arhoun et al. [23] reported different pH buffering processes that, while remaining active, can hide possible instabilities. Still, when the buffering capacity is depleted in the long term, abrupt pH changes could cause severe problems to the digester operation. However, other parameters can be an early warning for pH buffer depletion. Among the most used are total (TA), partial (PA), or intermediate (IA) alkalinity.

Alkalinity could be defined as the ability of the AD liquor of the mixture to buffer the possible generation of acids produced during the biological process and, hence, mitigate potential pH changes [7]. The alkalinity in the AD liquor mixture is mainly provided by the non-protonated forms of VFAs and the carbonate system. If no other species interfere within the pH range of anaerobic digesters, a VFAs accumulation would be directly related to the breakdown of the bicarbonate buffering capacity [25]. The PA, IA, and TA measurements can evaluate the relative buffering substances concentrations. TA measures the combined effect of different buffer systems and is calculated as TA = PA + IA. PA corresponds to the buffer capacity of the carbonate system, just as ammonium/ammonia. In contrast, IA is the difference between TA and PA and corresponds to the buffer capacity of the non-protonated forms of VFAs. Alkalinity titrated down to 5.75 pH value is defined as a PA, whereas TA is titrated to 4.30 [20, 26]. Some authors have reported 2000–4000 mg of $CaCO_3$ L^{-1} PA values as typical for properly performing digesters under mesophilic conditions and feeding organic solid waste [11, 27, 28]. However, in terms of stability, the evolution of parameters over time is more important than the actual concentration, acting as an early warning [23].

Several studies also reported alkalinity ratios as monitoring parameters used as early warning tools [23, 25, 29]. The process stability can be evaluated by the IA/PA ratio, which involves the acid concentration in the system (IA), and the buffer capacity provided by the carbonate species (PA). If the PA is insufficient to buffer the IA, the digester will be acidified, and the activity of microorganisms, especially methanogens, will be inhibited [29]. Therefore, to consider the process stable, the IA/PA ratio must be kept below 0.4. Some authors also indicate IA/TA ratio as a parameter for monitoring the anaerobic digestion process. However, this has lower sensitivity than the IA/PA ratio [25].

The VFAs are produced during the anaerobic degradation of organic solid waste, and their evolution provides information about the performance of the different AD steps [7, 15, 30]. Especially, the substrates with high biodegradabilities, such as fruit, vegetable, or food waste, have a higher tendency to generate VFAs. The most common VFAs are acetic (C2), propionic (C3), butyric (C4), and valeric (C5) [8]. Acetic acid has been described as the least toxic fatty acid. On the contrary, propionic acid concentration has been associated with system failure, being even more inhibitory than butyric acid [18, 31]. The propionic acid accumulation is probably related to its conversion being the least thermodynamically favorable [6]. According to some studies, a propionic acid concentration in the range of 0.45–3.00 g COD L^{-1} (COD, chemical oxygen demand) has a high potential to inhibit the process. Obviously, inhibition will also depend on the substrate treated and the operating conditions [6, 32, 33]. As a result, propionic acid is usually presented as the main parameter to follow when analyzing the stability of AD [7].

Another commonly used parameter to monitor the stability of the anaerobic digestion process is the ratio of volatile fatty acids (VFAs) to total alkalinity (TA) (VFA/TA or FOS/TAC ratio) [4, 34]. This parameter is related to the buffering capacity, represented by the total alkalinity, for a given effect of the VFA on the pH of the AD liquor mixture [23]. According to the literature, there are three critical VFA/TA ratio values. If VFA/TA ratio is lower than 0.40, the digester should be stable. When the ratio ranged from 0.40 to 0.80, the digester performance would present some signs of instability, while the VFA/TA ratio higher than 0.80 indicates significant instability in the digester [7, 11, 35].

2.2 Specific energy loading rate (SELR)

The specific energy loading rate (SELR) is, according to the literature, one of the parameters for evaluating the AD process stability, since it is useful for determining allowable organic loading rates. The SELR is defined as the quotient between the daily feed organic load (expressed in g of tCOD $(L{\cdot}d)^{-1}$) and the active biomass inside the digester (expressed in g VSS L^{-1}) (VSS, volatile suspended solids) (Eq. (1)) and can be considered as an indicator of food to mass ratio (F/M) [19]. Thus, if the food mass in the feedstock exceeds the mass of decomposer microorganisms, it could cause a metabolic imbalance because of the acidification and inhibition of methanogenic microorganisms [36]. On the contrary, if the abundance of food available is insufficient, the metabolism of the microorganisms could be affected [37]:

$$\text{SELR} = \frac{Q{\cdot}[tCOD]_{inlet}}{[VSS]{\cdot}V_{working}} \tag{1}$$

where Q is the inlet flow rate (L d^{-1}), $[tCOD]$ is feeding total COD concentration (g L^{-1}), $[VSS]$ is digestate volatile suspended solids concentrations (g L^{-1}), and V_{working} is the working volume of the digester (L).

According to Azevedo et al. [38], the limit value for SELR is 0.4 d^{-1}. A higher value indicates a potential instability between the biomass of the microbial consortium and the loading of the feed mixture.

2.3 Removal of organic matter (volatile solids and COD)

Methane production should be stable if there is no accumulation of organic compounds inside the digester. A feeding of substrates with poor biodegradability could increase the content of volatile solids inside the reactor. Likewise, if the feed rate exceeds the rate of degradation by the microorganisms, organic compounds will accumulate inside the reactor, and methane production will be impacted. It is worth noting that the biodegradability capacity of a digester would depend on substrate characteristics and microbial degradation capacity. Therefore, this variable would be useful to evaluate the adaptation of an AD reactor to new substrates by comparing the biodegradability values in the digester with the expected for the added substrates. In that sense, biomethane potential tests can be a powerful tool to provide a reference framework for the biodegradability of the substrates [39]. The volatile solid removal is determined by Eq. (2) [23, 35]:

$$VS_{\text{removal}} = \frac{VS_{inlet} \cdot VS_{digestate}}{VS_{inlet}} \tag{2}$$

2.4 Total ammonia nitrogen (TAN)

Feedstocks with high content of proteins, i.e. with high content of nitrogen compounds, could induce high total ammonia nitrogen (TAN) in the AD process leading to biomass inhibition [6, 35]. A C/N ratio of 10–30 in the feedstock has been reported in the literature, which could avoid ammonium inhibition [20, 23]. Ammonia inhibition usually leads to a decrease in methane production rate and an increase in intermediate VFAs. Ammonia levels in the 200–1000 mg NH_4-N L^{-1} have no adverse effect, while inhibition occurs between 1500 and 3000 mg NH_4-N L^{-1}, especially at higher pH values, and complete inhibition, at all pH values, above 3000 mg NH_4- N L^{-1} [35].

2.5 Biogas production and composition

Biogas production is a crucial measure of the AD process status. If at a given OLR, there is a decrease in biogas production or biogas production rate that does not correspond to the degradation of the fed load, it could be considered a warning sign that the process is not working at its optimum [20]. According to the literature, the production of biogas or methane can be expressed as gas production rate (GPR), specific gas production (SGP), specific methane production (SMP), or specific methane yield (SMY), among others [6, 38, 40]. The GPR is expressed as the biogas volume generated per day to the reactor volume. The SGP and SMP/SMY are the biogas or methane volume generated by the mass of volatile solids feeding [6]. Generally, specific parameters are used to compare the stability of anaerobic digestion processes developed at different OLR values.

The organic matter degradation by microorganisms produces different types of gasses contained in the biogas. The biogas composition is mainly methane (45–85%), carbon dioxide (15–45%), and other gases such as hydrogen sulfide, ammonia, and nitrogen. The accurate proportion of gasses in the biogas depends on the process conditions and the feedstock [20].

A change in the microbial community could generate a different biogas composition. If there is an accumulation of hydrogen in the process, the hydrogenotrophic methanogenic microorganisms could be inhibited. On the contrary, if acetic acid accumulates, the acetoclastic methanogenic microorganisms could be inhibited. In both situations, methane production could be affected [18]. Also, the different compositions of the feedstock can directly affect biogas generation. Alibardi & Cossu [41] found that a higher proportion of carbohydrates in the substrate results in a more significant biogas generation. Not so when the substrate is mainly composed of lipids or proteins.

2.6 Microbial population adaptation

Commonly, the AD process involves several stages: hydrolysis, acidogenesis, acetogenesis, and methanogenesis [20, 21]. First, hydrolytic microorganisms are responsible for breaking down complex organic matter, such as proteins and carbohydrates, into simpler compounds. In the acidogenesis stage, the simpler compounds are biodegraded by acidogenic into VFAs, alcohols, H_2, and CO_2. Then, acetogenic microorganisms transform them into acetic acid, H_2, and CO_2. Finally, methanogens convert these products into CH_4 and CO_2 [42] following two pathways. One is carried out by acetoclastic methanogens, which can convert acetate into CH_4 and CO_2; the other is performed by hydrogenotrophic methanogens, which convert H_2/CO_2 to CH_4 [43]. These relationships play a crucial role in the anaerobic process leading to a balance between populations. For example, hydrogenotrophic methanogens are responsible for maintaining a low partial pressure of H_2 (<10 Pa), which is necessary for the functioning of the intermediate trophic group [44]. Therefore, the AD process stages efficiency is closely associated with the abundance and activities of specific anaerobic microbial communities. Many studies have reported that the diversity and abundance of microbial communities are closely associated with the digestion conditions, such as OLR, pH, temperature, HRT, and types of digestion substrates [2, 21], and an active anaerobic microbial communities imbalance could reduce the efficiency of the AD process [21, 45].

In anaerobic digesters, the stability of the microbial population and the relationships between groups (i.e. acetate utilizing methanogens/hydrogen utilizing methanogens ratio, and sulfate-reducing bacteria/methanogens ratio) are widely used parameters to establish the stability of the digesters. However, there is a lack of studies reporting the effect of feedstock's type and composition on community structure and microbial activity changes. Zahedi et al. [42] have observed that although the number of microorganisms is essential in many microbial ecology studies of anaerobic digestion, operating with actual and changing wastes under realistic circumstances, MSW could be not a key parameter to control the process. It was concluded that stability and good microbial community dynamics (flexibility to adapt in response to changes in environments, particularly to changes in the substrate and operating conditions) are essential factors for the stable performance of the reactors [42].

So, although some researchers have documented that feedstock composition and OLR may influence bacterial and archaeal communities, there is a lack of consensus on the impact assessment that drastic changes in the type and composition of feedstock might have on community structure changes, bacterial density increases, and microbial diversity. Furthermore, nowadays, the complexity, cost, and high expertise required for this kind of analysis advocate for using the microbial analyses as a supplementary tool for gaining deep knowledge of the reactors, but not for the routine monitoring of the AD process.

3. Feedstock changes in the anaerobic digestion process of organic solid waste

This section reports information compiled from the literature on studies investigating the transitory states during the AD process when the type or composition of the feedstock fed has changed. A change in feedstock type would refer to a feed with different substrates, whereas a change in feedstock composition would refer to a feed where the

Feedstock	Feeding	Type of AD	Monitoring parameters	Temp.	OLR	Methane production	Ref.
Cow manure (CW), Sugar beet pulp (SBP), Linen (Ln), and Wheat straw (WS)	CW/Ln/WS CW/Ln/SBM	AcoD	Biogas production Biogas composition % VS removal	Mesophilic (37 ± 1°C)	1.0 g VS $(L{\cdot}d)^{-1}$	CW/Ln/WS: 0.064 L g VS^{-1} CW/Ln/SBM: 0.191 L g VS^{-1}	[46]
Fruit pulp waste by a fruit juice company	Peach Raspberry White guava	AmD (Two steps)	VFAs Biogas production Biogas composition Microbiology %COD removal	Mesophilic (30°C)	7.0 to 25.7 g COD $(L{\cdot}d)^{-1}$ (acidogenic reactor) 1.9 to 7.4 g COD $(L{\cdot}d)^{-1}$ (methanogenic reactor)	Peach: 0.30 L g COD^{-1} Raspberry-I: 0.30 L g COD^{-1} Raspberry-II: 0.32 L g COD^{-1} White guava: 0.37 L g COD^{-1}	[47]
Municipal sewage sludge (SS) and orange peel (OP) from a bar	Stage I: SS Stage II: SS+OP pre-reated (OL) Stage III: SS+Sieved OL (SOL)	AmD and AcoD	pH Biogas production Biogas composition SELR	Mesophilic (37 ± 2°C)	1.80 ± 0.31 g VS $(L{\cdot}d)^{-1}$	SS: 0.100 L g VS^{-1} SS+OL: 0.177 L g VS^{-1} SS+SOL: 0.301 L g VS^{-1}	[40]
Fruit pulp waste by a juice-producing company	Peach Apple	AmD (Two steps / lab scale)	VFAs Biogas production Biogas composition Microbiology %COD removal	Mesophilic (30–37°C)	21.2 to 51.1 g COD $(L{\cdot}d)^{-1}$ (acidogenic reactor) 0.2 to 12.2 g COD $(L{\cdot}d)^{-1}$ (methanogenic reactor)	Peach: 0.25 L g COD^{-1} Apple: 0.31 L g COD^{-1}	[45]
	Pear Apple	AmD (Two steps / pilot scale)		Mesophilic (30°C)	12.2 to 22.2 g COD $(L{\cdot}d)^{-1}$ (acidogenic reactor) 1.8 to 3.3 g COD $(L{\cdot}d)^{-1}$ (methanogenic reactor)	Pear: 0.30 L g COD^{-1} Apple: 0.30 L g COD^{-1}	

Feedstock	Feeding	Type of AD	Monitoring parameters	Temp.	OLR	Methane production	Ref.
Artificial organic fraction (AOF) and four agro-industrial wastes (AWs) (cotton gin waste (CGW), winery waste (WW), olive pomace (OP), and juice industry waste (JW)) *Assays A: single feedstock Assays B: co-feedstock in a ratio of 40:60 in VS AW: AOF	A-I: CGW-WW-OP-JW-CGW A-II: WW-OP-JW-CGW-WW A-III: OP-JW-CGW-WW-OP A-IV: JW-CGW-WW-OP-JW	AmD	pH Alkalinity VFAs VFA/TA ratio TAN Biogas production	Mesophilic (35°C)	1.0 g VS (L·d)$^{-1}$	A-I: 0.201 → 0.148 L g VS^{-1} A-II: 0.297 → 0.338 L g VS^{-1} A-III: 0.117 → 0.151 L g VS^{-1} A-IV: 0.335 → 0.369 L g VS^{-1}	[6]
	B-I: CGW-WW-OP-JW-CGW B-II: WW-OP-JW-CGW-WW B-III: OP-JW-CGW-WW-OP B-IV: JW-CGW-WW-OP-JW	AcoD				B-I: 0.280 → 0.284 L g VS^{-1} B-II: 0.384 → 0.405 L g VS^{-1} B-III: 0.268→ 0.319 L g VS^{-1} B-IV: 0.402 → 0.429 L g VS^{-1}	
Sewage sludge (SS) from a municipal WWTP and fruit waste from a fruit processing industry (peach waste (PW), banana waste (BW), and apple waste (AW))	Stage I: SS Stage II: SS + PW Stage III: SS + BW Stage IV: SS + AW Stage V: SS	AmD and AcoD	Alkalinity VFAs VFA/TA ratio Biogas production Biogas composition	Mesophilic (37 ± 2°C)	1.2 g VS (L·d)$^{-1}$ (AmD) 3.0 g VS (L·d)$^{-1}$ (AcoD)	SS: 0.28 L g VS^{-1} SS+PW: 0.20 L g VS^{-1} SS+BW: 0.30 L g VS^{-1} SS+AW: 0.26 L g VS^{-1} SS: 0.28 L g VS^{-1}	[7]

AmD: anaerobic mono-digestion; AcoD: anaerobic co-digestion, where OLR, organic loading rate; VFA, volatile fatty acids; TA, total alkalinity; VS, volatile solids; TAN, total ammonia nitrogen; SELR, specific energy loading rate; and COD, chemical oxygen demand.

Table 1.
Type of feedstock change.

percentage composition of the various substrates that compose the feedstock varies. In these works, to evaluate whether the AD process was finally able to adapt to the perturbations of the system, the parameters previously described in section 2 were assessed.

3.1 Type of feedstock

This section includes all the studies found in the literature that evaluate the AD process stability when changing the feedstock type (**Table 1**). These studies mostly use seasonal wastes generated in agri-food industries, i.e. fruit or vegetable processing, as feedstock, such as fruit pulp waste by a juice-producing company, winery waste, olive pomace, or sugar beet pulp. All these studies agree that the seasonality of the fruit and vegetable processing industries and waste from different crops would complicate operating a digester under the same conditions over a long period, because the waste supply could be changed or discontinued frequently [7, 46]. Therefore, using a single digester, fed with multiple feedstocks generated in the same geographical area and strongly dependent on seasonality, would require a deep knowledge of the behavior of the AD process when exposed to the resulting feed changes.

Despite the limited literature on the field, there is a wide variety of approaches for assessing the effect of feedstock type change on the stability of the AD process. Feedstock type change has been evaluated in mono-digestion processes with sequential feeding [6] or two-stage processes [45, 47]. It has also been studied in the transition from mono-digestion to co-digestion by applying feedstock change in the latter case [7, 40] and co-digestion processes with sequential feeding [6] or multi-substrate [46].

All research that has monitored pH as a stability parameter has used substrates from similar origins and characteristics, so it has reported stable pH values between 7.0 and 8.0 [6, 40]. As for monitoring alkalinity, VFAs concentration, and VFA/TA ratio, variable results have been reported, all of them related to the varying composition of the feedstocks fed. According to Pellera et al. [6], who evaluated the sequential feeding of four agro-industrial feedstocks (CGW → WW → OP → JW; cotton gin was (CGW), juice industry waste (JW), olive pomace (OP), and winery waste (WW)), the VFAs concentration was higher during the first two stages, especially for the experiments that started with feeding the most biodegradable feedstocks, i.e. WW and JW. Then, the values decreased to stable levels until the end of the experimentation, while the TA showed an increasing trend. Similarly, the VFA/TA ratio followed the same trend that VFA, without exceeding the value of 0.4, thus corroborating the system's stability. In contrast, Fonoll et al. [7] stated that feedstock changes did not increase VFAs concentration. However, due to the different biodegradability of fruit wastes, methane production and digester alkalinity changed to a lesser extent. The VFA/TA ratio values showed stability while changing feedstock despite the observed alkalinity fluctuations. Carvalheira et al. [45] evaluated the feedstock change in a two-stage anaerobic mono-digestion process, using fruit pulp waste by a juice-producing company as a substrate. During the monitoring of the acidogenic reactor, differences in the profile of fermentation products, i.e. VFAs, lactic acid, and ethanol, were identified and quantified when using other fruit pulp wastes. These results were attributed to carbohydrate concentration and OLR on the effluent composition. On the contrary, Mateus et al. [47], who also evaluated the feedstock change in a two-stage anaerobic mono-digestion process, reported a stable fermentation product profile, regardless of the different carbohydrate concentrations in the substrates and OLR changes. The difference between both studies could be attributed to the fact that the OLR range used by Carvalheira et al. [45] to apply the feedstock change was higher.

The evaluation of AD process stability when feedstock type changes through SELR was reported by Carvalho et al. [40]. The SELR values ranged between 0.22 and 0.33 d^{-1}, without significant differences, keeping the values below 0.4 d^{-1} and ensuring that the digestor worked under stable conditions (section 2.2).

Concerning methane production and composition, available research reports stable production values and relates their differences to the characteristics and biodegradability of the feedstocks fed [6, 7, 40, 46]. However, one of the most remarkable results dealing with methane production was reported in the study by Pellera et al. [6], which evaluated a sequential feeding by mono-digestion and co-digestion. Methane production with the same feedstock fed in different feeding sequences had similar values, attributed to an immediate response of the microbial population to each substrate. In fact, after providing the digesters with four feedstocks (mono- or co-substrate) in sequential order, the last feeding was carried out with the feedstock that had been fed first in each assay. The results demonstrated that the final methane production values were higher than their first values in all cases (**Table 1**). As an explanation for these results, they suggested a positive level of microbial population adaptation, albeit also possible presence of higher amounts of degradable material in the reactors as it was fed on 14 times. On the other hand, Carvalheira et al. [45] showed an increase in fermentation product concentration in the effluent of the methanogenic reactor, with the change of substrate reaching a maximum of 6.565 g COD L^{-1}, even after decreasing the OLR. There was a significant acetic and propionic acid accumulation, 2.44 and 1.44 g COD L^{-1}, respectively. The decrease in OLR and biodegradable matter accumulation decreased methane production when peach pulp was replaced by apple pulp, from 4.33 to 3.38 g COD $(L{\cdot}d)^{-1}$, respectively (**Table 1**). The decrease in process efficiency indicated that the microbial community was affected by the influent change and could not treat the apple influent with a high OLR as efficiently as the previous peach influent. Despite influent variations, stable performance of the methanogenic stage was achieved, probably due to the buffering capacity of the acidogenic community at the initial stage. In contrast, Mateus et al. [47] reported differences in the biogas composition generated in the acidogenic step in evaluating the two-stage mono-digestion process. In this case, the difference in carbohydrate concentration seemed to mainly affect the gas production and composition in the acidogenic reactor. No hydrogen production was detected with the peach pulp waste but with the raspberry and white guava pulps waste, ranging from 4 to 34%.

Reviewed studies state that whether or not there was instability during the whole AD experiment when the feedstock type changes, the microbial population has acclimatized well to the change. Different authors have argued that an acclimatization period would not be necessary with each change of feed material, as the microbial community is already adapted to substrates of a similar nature. Studies assessing changes in the microbial population ensure that the reactors were abundant in archaeal methanogens, mainly *Methanosaeta*, responsible for acetoclastic methanogenesis, the most common process in AD processes involving the production of CH_4 and CO_2 from acetate. *Methanobacterium*, microorganisms responsible for hydrogenotrophic methanogenesis involving methane production from CO_2 and H_2, were also identified. The microbial community composition remained relatively constant over time in each experiment [45, 47].

3.2 Composition of feedstock

This section includes all the studies found in the literature that evaluate the AD process stability when changing the feedstock composition in the influent (**Table 2**). These studies mostly use a mixture of wastes whose composition is strongly dependent on seasonality, such as food waste, fruit and vegetable waste from wholesale markets, meat waste, or the organic fraction of municipal solid waste (OFMSW).

All these studies aimed to evaluate changes in feedstock composition in the influent on the digesters' stability. However, some assays kept the organic loading rate (OLR) constant throughout the experimentation, despite the change in influent composition, to attribute the changes in reactor behavior to the change in composition [48, 49]. On the contrary, in other studies, by ignoring the intrinsic modification of the OLR due to the change in composition due to percentage (w:w or v:v) increase, they evaluated the combined effect of these two factors [2, 4, 23, 26, 34, 35, 43].

Unlike described in section 3.1, despite the limited literature in this field, not many different approaches have been studied to assess the effect of changing feedstock composition on the stability of the AD process. Feedstock composition change has been evaluated in the single- and two-stage mono-digestion process at the pilot scale [4, 34] and in the transition from mono-digestion to co-digestion by increasing the co-substrate percentage in the feed mixture [2, 23, 26, 35, 43, 48, 49]. Some studies have implemented changes in compositional percentages to improve methane production by adjusting the C/N ratio and the most optimal fruit and vegetable percentage.

Some research that monitored pH as a stability parameter by changing the feedstock composition in the influent has reported stable pH values between 7.0 and 8.0 and within the optimal range described in the literature for methanogenic bacteria [2, 26, 35, 43]. However, some other studies have reported fluctuations in these parameters. For example, Arhoun et al. [23], who evaluated the change in feedstock composition and seasonal variations, observed a very slight trend of decreasing pH with winter substrate. This slight acidification was related to the mixture's pH value, which was 3.5, lower than the other seasons, approximately 4.8. Masebinu et al. [4] and Scano et al. [34] have also reported a slight decrease in pH values with an increasing percentage of fruit in the feedstock composition, a higher percentage of citrus fruit, and fruits with a very high content of simple sugars, respectively. In addition, García-Peña et al. [49] have also described a quick drop in pH when feeding the reactors with FVW that was solved by adding buffer (NaOH 0.8 M) to supply the appropriate buffering capacity and avoid excessive pH drop under unbalanced conditions.

As indicated in section 3.1, regarding the monitoring of alkalinity, VFA concentration, and VFA/TA ratio, variable results have been reported, all of them related to the varying composition of the feedstocks fed and the specific stress situations performed during reactor feeding.

Some authors assessing alkalinity and VFAs and their corresponding ratios report stable values, and compliance with stability recommendations for VFA/TA and IA/PA ratios has been reported in the literature [23, 26, 35]. Tonanzi et al. [2] have reported a slight transient accumulation of acetic acid (60% of the soluble content) as a result of an increase in OLR up to 3.5 g VS $(L\ d)^{-1}$, reflected in a decrease in methane production. Propionic acid remained at low levels. The robustness of the microbiome and buffering capacities ensured quick recovery, acetic acid was eliminated, and methane production reached a stable value of 0.29 NL^3 CH^4 g VS^{-1} (**Table 2**). Masebinu et al. [4] have observed two significant instabilities caused by reaching high OLRs (3.42 and 4.06 g VS $(L\ d)^{-1}$), i.e. mixtures with high fruit concentrations. A high OLR causes the system to be susceptible to fluctuations in feed composition and operating parameters. Maintaining a high fruit fraction in the substrate mixture caused a decrease in pH, an increase in the VFA/TA ratio above the stable region (0.45 and 0.53), and an eventual reduction in biogas production. As the percentages of fruit in the feed mix were reduced, all improved and returned to stability with improved biogas yield. Both pH and VFA/TA immediately indicated instability for an exceptionally high fruit concentration.

Feedstock	Feeding	Type of AD	Monitoring parameters	Temp.	OLR	Methane production	Ref.
Waste activated sludge (WAS) and food waste (FW) (FW:WAS ratio, on a VS basis)	Stage A-I: 100:0 Stage A-II: No feed Stage A-III: 100:0 Stage A-IV: No feed Stage A-V: 30:70	AmD and AcoD	pH VFAs Biogas production Microbiology	Mesophilic (37°C)	0.8–3.5 g VS $(L{\cdot}d)^{-1}$	A-V: 0.17 NL g VS^{-1}	[2]
	Stage B-I: 70:30 Stage B-II: 90:10 Stage B-III: 95:5 Stage B-IV: 100:0	AmD and AcoD				B-I: 0.27 NL g VS^{-1} B-II: 0.29 NL g VS^{-1} B-III: 0.29 NL g VS^{-1} B-IV: 0.23 NL g VS^{-1}	
Food waste (FW) and sewage sludge (SeS) (FW:SeS ratio, TS %)	Stage I: 100:0 Stage II: 100:0 Stage III: 75:25 Stage IV: 50:50 Stage V: 25:75 Stage VI: 0:100	AmD and AcoD	pH Biogas production Biogas composition Microbiology	Mesophilic (37 ± 2°C)	1.41 – 3.30 g VS $(L{\cdot}d)^{-1}$	I: 0.54 L g VS^{-1} II: 0.54 L g VS^{-1} III: 0.46 L g VS^{-1} IV: 0.37 L g VS^{-1} V: 0.31 L g VS^{-1} VI: 0.30 L g VS^{-1}	[43]

Feedstock	Feeding	Type of AD	Monitoring parameters	Temp.	OLR	Methane production	Ref.
Mixed sewage sludge (MSS) and fruit and vegetable waste (FVW) by a wholesale market (FVW:MSS ratio, based on (v/v)) (S: summer; A: autumn; W: winter and Sp: spring)	Stage S-I: 0:100 Stage S-II: 20:80 Stage S-III: 40:60 Stage S-IV: 60:40 Stage S-V: 80:20 Stage S-VI: 100:0	AmD and AcoD	pH Alkalinity IA/PA ratio IA/TA ratio Biogas production Biogas composition % VS removal	Mesophilic (35°C)	0.6 to 5.5 g VS $(L{\cdot}d)^{-1}$	S-I: 0.276 L g VS^{-1} S-II: 0.346 L g VS^{-1} S-III: 0.33 L g VS^{-1} S-IV: 0.355 L g VS^{-1} S-V: 0.362 L g VS^{-1} S-VI: 0.323 L g VS^{-1}	[23]
	Stage A-I: 0:100 Stage A-II: 20:80 Stage A-III: 40:60 Stage A-IV: 60:40 Stage A-V: 80:20 Stage A-VI: 100:0				1.0 to 4.8 g VS $(L{\cdot}d)^{-1}$	A-I: 0.308 L g VS^{-1} A-II: 0.320 L g VS^{-1} A-III: 0.441 L g VS^{-1} A-IV: 0.447 L g VS^{-1} A-V: 0.409 L g VS^{-1} A-VI: 0.416 L g VS^{-1}	
	Stage W-I: 0:100 Stage W-II: 20:80 Stage W-III: 40:60 Stage W-IV: 60:40 Stage W-V: 80:20 Stage W-VI: 100:0				1.8 to 5.8 g VS $(L{\cdot}d)^{-1}$	W-I: 0.329 L g VS^{-1} W-II: 0.371 L g VS^{-1} W-III: 0.380 L g VS^{-1} W-IV: 0.418 L g VS^{-1} W-V: 0.405 L g VS^{-1} W-VI: 0.393 L g VS^{-1}	
	Stage Sp-I: 0:100 Stage Sp-II: 20:80 Stage Sp-III: 40:60 Stage Sp-IV: 60:40 Stage Sp-V: 80:20 Stage Sp-VI: 100:0				0.6 to 4.8 g VS $(L{\cdot}d)^{-1}$	Sp-I: 0.323 L g VS^{-1} Sp-II: 0.345 L g VS^{-1} Sp-III: 0.325 L g VS^{-1} Sp-IV: 0.340 L g VS^{-1} Sp-V: 0.348 L g VS^{-1} Sp-VI: 0.394 L g VS^{-1}	
Mixed sewage sludge (MSS) and fruit and vegetable waste (FVW) by a wholesale market (FVW:MSS ratio, based on (v/v))	Stage I: 0:100 Stage II: 20:80 Stage III: 40:60 Stage IV: 60:40 Stage V: 80:20 Stage VI: 100:0	AmD and AcoD	pH Alkalinity IA/PA ratio Biogas production Biogas composition % VS removal	Mesophilic (35 ± 1°C)	1.03 to 4.78 g VS $(L{\cdot}d)^{-1}$	I: 0.303 L g VS^{-1} II: 0.380 L g VS^{-1} III: 0.445 L g VS^{-1} IV: 0.405 L g VS^{-1} V: 0.390 L g VS^{-1} VI: 0.403 L g VS^{-1}	[26]

Feedstock	Feeding	Type of AD	Monitoring parameters	Temp.	OLR	Methane production	Ref.
Fruit and vegetable waste by a wholesale market (Fruit:Vegetable ratio fruit fraction %)	FVWs collected weekly with different % of fruit and vegetables. Cycle I: 36.89% fruit Cycle II: 52.83% fruit Cycle III: 44.62% fruit	AmD (Two steps/ pilot scale)	pH Alkalinity VFAs VFA/TA ratio Biogas production Biogas composition % VS removal	Mesophilic (35 ± 1°C)	0.5 to 4.06 g VS $(L{\cdot}d)^{-1}$	I: 0.35 NL g VS^{-1} II: 0.51 NL g VS^{-1} III: 0.55 NL g VS^{-1}	[4]
Source selected of organic fraction of municipal solid waste (SS-OFMSW) collected from a canteen and in a fruit and vegetable wholesale market, sewage sludge (SwS) from a WWTP thickener and treated wastewater (TW) (SwS:SS-OFMSW:TW ratios, weight-based)	Stage I: 100:0:0 Stage II: 90.9:1.5:7.6 Stage III: 90.9:3.0:6.1 Stage IV: 66.7:11.1:22.2 Stage V: 66.7:16.7:16.6 Stage VI: 41.3:29.3:29.4	AmD and AcoD	pH Alkalinity VFAs TAN Biogas production Biogas composition % VS removal % COD removal	Mesophilic (37–38°C)	I: 0.80 g VS $(L{\cdot}d)^{-1}$ II: 1.10 g VS $(L{\cdot}d)^{-1}$ III: 0.94 g VS $(L{\cdot}d)^{-1}$ IV: 1.23 kg VS $(m^3\ d)^{-1}$ V: 1.74 kg VS $(m^3\ d)^{-1}$ VI: 3.20 kg VS $(m^3\ d)^{-1}$	I: 0.25 NL g VS^{-1} II: 0.26 NL g VS^{-1} III: 0.32 NL g VS^{-1} IV: 0.32 NL g VS^{-1} V: 0.37 NL biogas g VS^{-1} VI: 0.49 NL biogas g VS^{-1}	[35]
Real digester effluente (RW from residual waste), waste paper plus cardboard (WP) and biowaste (BioW) diluted with digester supernatant from the plant (RW:BioW:WP ratios, based on (w:w))	Stage I: 100:0:0 Stage II: 0:100:0 Stage III: 0:85:15 Stage IV: 0:70:30	AmD and AcoD	Alkalinity VFAs VFA/TA ratio Biogas production Biogas composition % VS removal	Mesophilic (35°C)	2.9 ± 0.4 g VS $(L_R\ d)^{-1}$	I: 0.34 L g VS^{-1} II: 0.41 L g VS^{-1} III: 0.36 L g VS^{-1} IV: 0.34 L g VS^{-1}	[48]

Feedstock	Feeding	Type of AD	Monitoring parameters	Temp.	OLR	Methane production	Ref.
Fruit and vegetable waste (FVW) by a wholesale market (fruit:vegetable ratio, fruit fraction %)	Different fruit and vegetable mixtures with a fruit fraction of 33.4, 33.1, 28.4 and 60.6%	AmD (pilot scale)	pH Alkalinity VFAs VFA/TA ratio Biogas production Biogas composition % VS removal	Mesophilic (35 ± 0.5°C)	0.5 to 5.0 kg VS $(m^3\ d)^{-1}$	I: 0.47 NL g VS^{-1} II: 0.39 NL g VS^{-1} III: 0.44 NL g VS^{-1} IV: 0.46 NL g VS^{-1}	[34]
Fruit and vegetable waste from the central food distribution market and meat residues (MR) (FVW;MR ratios, based on (v/v))	Stage I: 100:0 Stage II: 75:25 Stage III: 50:50 Stage IV: 100:0 Stage V: 50:50 Stage VI: 75:25	AmD and AcoD	pH VFAs TAN Biogas production Biogas composition Microbiology % VS removal	Room temperature	2.4 to 2.7 g COD $(L\ d)^{-1}$	I: 0.10 L g VS^{-1} II: 0.14 L g VS^{-1} III: 0.12 L g VS^{-1} IV: 0.04 L g VS^{-1} V: 0.04 L g VS^{-1} VI: 0.14 L g VS^{-1}	[49]

AmD: anaerobic mono-digestion; AcoD: anaerobic co-digestion), where OLR, organic loading rate; VFA, volatile fatty acids; IA, intermediate alkalinity; PA, partial alkalinity; TA, total alkalinity; VS, volatile solids; TS, total solids; TAN, total ammonia nitrogen; and COD, chemical oxygen demand.

Table 2.
Composition of feedstock change.

On the other hand, in the research carried out by Fonoll et al. [48] and Scano et al. [34], the digestion systems were subjected to stressful scenarios to compare the robustness of the process with respect periods of stability. Fonoll et al. [48] have reported that 15% and 30% replacement of biowaste (BioW) with waste paper (WP) did not affect VFA and alkalinity levels. However, for a replacement of 30%, acidification of the supernatant used to dilute the feedstock led to a rapid accumulation of VFA (2400 mg L^{-1}), which decreased methane production. Recovery was carried out after a period without feeding and by re-establishing the feed supply using a new batch of supernatant. Scano et al. [34] observed an initial increase in VFA/TA, reaching values close to 0.65, corresponding to the increase in OLR due to a high percentage of fruit in the feed mixture. As a corrective strategy, the percentage of fruit in the mix was reduced, resulting in a corresponding reduction in VFA/TA. During the subsequent stages, VFA/TA was mainly influenced by chemical composition differences of the feed substrate, changes in OLR, and the simple sugar content of the fruit waste. Experimental results reported that the AD process still performed well with well-balanced mixtures of fruit and vegetable wastes, even for VFA/TA of up to 0.5. The highly elevated VFA/TA values (above 1) were derived from specific stress tests performed by feeding the reactor with substantial quantities of substrates with high content of simple sugars. The substrate mixture's large melon (which contains large amounts of highly degradable sugars) caused significant instability. Furthermore, it was complicated to stabilize the process at the next stage, as the available VWFs were mainly composed of fruit waste.

The assessment of the stability of the AD process when the feedstock composition changes through the TAN concentration measurement was reported by Cabbai et al. [35]. Organic nitrogen from the feed substrates (SS-OFSMW and SwS) was metabolized by the biomass-producing ammonia, although the levels were safe for process stability. García-Peña et al. [49], who evaluated the co-digestion of FVW by varying the percentage of meat residue (MR), reported that the addition of MT (75:25), rich in protein, started to release ammonia from the hydrolysis of the protein, which favored an increase in the alkalinity of the medium and the pH drop regulation.

Regarding methane production and composition, all the researchers have reported stable production values and related their differences to the characteristics and biodegradability of the feedstocks fed, just as to the different OLRs evaluated (**Table 2**).

In cases of feedstock composition change evaluating the microbial population adaptation, changes have been observed in contrast to feedstock-type changes. Tonanzi et al. [2] and Cheng et al. [43], who evaluated co-digestion of activated sewage sludge with percentage changes of FW, detected hydrolytic bacteria growth, such as *Bacteroidales*, especially the *Prolixibacteriaceae* family, whose relative abundance increased linearly with FW percentage in the mixture composition. As for the archaeal populations, high diversity indices were found when the FW and activated sewage sludge percentages in the feedstock composition varied, suggesting that the archaeal biodiversity was affected by the reactor feed conditions. Most of the acetoclastic methanogens determined belonged to the order *Methanosarcinales*, mainly to the genus *Methanosaeta*. In contrast, the hydrogenotrophic methanogens identified belonged to the orders *Methanomicrobiales* and *Methanobacteriales*, mainly to the genus *Methanobacterium*. The combined relative abundances of the three methanogens did not show significant changes in the two investigations. However, it was clear that *Methanosaeta* competed over *Methanobacterium*, and the latter had advantages with increasing FW in the feedstock composition. Furthermore, Tonanzi et al. [2] stated that minimal activated sewage sludge addition (FW: WAS, 95:5) enriched the microbial community with *Methanospirilloun* and *Candidatus Methanophastidiosun*, which have a high H_2-consuming capacity, avoiding thermodynamic bottlenecks and failure of the FW mono-digestion process by the drop in activity of acetoclastic microorganisms as a result of a dramatic accumulation of propionic acid.

On the other hand, García-Peña et al. [49] reported enrichment of the microbial population of Firmicutes (fermenting bacteria that degrade VFAs), just to Bacteroidetes (proteolytic bacteria probably involved in the degradation of MR) when the percentage of MR increased in the composition of the raw material feeding. As for the archaea population, it was reported that the hydrogenotrophic methanogenic genus *Methanobacteriales* represented more than 93% of the presence of archaea in the digester. The hydrogenotrophic methanogenic community dominated even though the digester was inoculated with cow manure, which commonly contains acetoclastic methanogens. Finally, it may be argued that when changes are carried out to the composition of the feedstock fed with substrates of very diverse origins and biodegradability, microbial populations adapt to the changes in the end.

4. Conclusions

Different control strategies and parameters are reported in the literature to monitor system stability as a result of changes related to the feedstock. Several strategies have been implemented, reactor operating time extensions, lowering of OLRs, or supplementing enough alkalinity to buffer possible pH shocks. A longer operation time could allow the choice of a transition strategy to microbial community adaptation. Also, systems operation with low OLRs is less disturbed by instabilities. In any case, the transitory state is when the process needs to be carefully controlled not to reach a point of irreversible inhibition. The control parameters that should be monitored to prevent this situation would be pH and VFA/TA ratio. The microbial population analysis provides interesting information on its adaptation, although it does not reveal enough details for daily monitoring. Developing protocols for substrate changeover strategies could be of great interest to expand the use of existing anaerobic digestion facilities. It would improve the potential of this technology in the treatment of seasonal substrates that are generally not treated by anaerobic digestion. Concerning anaerobic digestion process robustness, changes in the type or composition of feedstock fed to the reactors are possible, although cautiously.

Acknowledgements

The authors acknowledge support of the publication fee by the CSIC Open Access Publication Support Initiative through its Unit of Information Resources for Research (URICI). The authors Ángeles Trujillo-Reyes and Dr. Fernando G. Fermoso would like to thank the project entitled “Employing circular economy approach for OFMSW management within the Mediterranean countries-CEOMED” number A_B4.2_0058, funded under the ENI CBC MED 2014-2020 program, for financing this research. Dr. Antonio Serrano is grateful to the Economic Transformation, Industry, Knowledge, and Universities Department of the Andalucia Autonomous Government for this Emergia fellowship (EMERGIA20_00114). Dr. David Jeison would like to thank the support provided by CRHIAM centre (ANID/FONDAP/15130015).

Conflict of interest

The authors declare no conflict of interest.

Author details

Ángeles Trujillo-Reyes[1], Sofía G. Cuéllar[2], David Jeison[2], Antonio Serrano[3,4], Soraya Zahedi[1] and Fernando G. Fermoso[1*]

1 Instituto de la Grasa, Spanish National Research Council, Seville, Spain

2 Escuela de Ingeniería Bioquímica, Pontificia Universidad Católica de Valparaíso, Valparaíso, Chile

3 Institute of Water Research, University of Granada, Granada, Spain

4 Pharmacy Faculty, Department of Microbiology, University of Granada, Granada, Spain

*Address all correspondence to: fgfermoso@ig.csic.es

References

[1] Wilson DC, Rodic L, Modak P, Soos R, Carpintero Rogero A, Velis C, et al. Global Waste Management Outlook: United Nations Environment Programme. Vienna, Austria: International Solid Waste Association; 2015

[2] Tonanzi B, Gallipoli A, Gianico A, Montecchio D, Pagliaccia P, Rossetti S, et al. Elucidating the key factors in semicontinuous anaerobic digestion of urban biowaste: The crucial role of sludge addition in process stability, microbial community enrichment and methane production. Renewable Energy. 2021;**179**:272-284

[3] Kaza S, Yao L, Bhada-Tata P, Van Woerden F. What a Waste 2.0: A Global Snapshot of Solid Waste Management to 2050. Washington, USA: World Bank Publications; p. 2018

[4] Masebinu SO, Akinlabi ET, Muzenda E, Aboyade AO, Mbohwa C. Experimental and feasibility assessment of biogas production by anaerobic digestion of fruit and vegetable waste from Joburg Market. Waste Management. 2018;**75**:236-250

[5] Mirmohamadsadeghi S, Karimi K, Tabatabaei M, Aghbashlo M. Biogas production from food wastes: A review on recent developments and future perspectives. Bioresource Technology Reports. 2019;**7**:100202

[6] Pellera F-M, Gidarakos E. Anaerobic digestion of solid agroindustrial waste in semi-continuous mode: Evaluation of mono-digestion and co-digestion systems. Waste Management. 2017;**68**:103-119

[7] Fonoll X, Astals S, Dosta J, Mata-Alvarez J. Anaerobic co-digestion of sewage sludge and fruit wastes: Evaluation of the transitory states when the co-substrate is changed. Chemical Engineering Journal. 2015;**262**:1268-1274

[8] Wainaina S, Awasthi MK, Sarsaiya S, Chen H, Singh E, Kumar A, et al. Resource recovery and circular economy from organic solid waste using aerobic and anaerobic digestion technologies. Bioresource Technology. 2020;**301**:122778

[9] Bolzonella D, Battistoni P, Mata-Alvarez J, Cecchi F. Anaerobic digestion of organic solid wastes: Process behaviour in transient conditions. Water Science and Technology. 2003;**48**(4):1-8

[10] Bolzonella D, Pavan P, Battistoni P, Cecchi F. Anaerobic co-digestion of sludge with other organic wastes and phosphorus reclamation in wastewater treatment plants for biological nutrients removal. Water Science and Technology. 2006;**53**(12):177-186

[11] Bouallagui H, Lahdheb H, Romdan EB, Rachdi B, Hamdi M. Improvement of fruit and vegetable waste anaerobic digestion performance and stability with co-substrates addition. Journal of Environmental Management. 2009;**90**(5):1844-1849

[12] Boušková A, Dohanyos M, Schmidt JE, Angelidaki I. Strategies for changing temperature from mesophilic to thermophilic conditions in anaerobic CSTR reactors treating sewage sludge. Water Research. 2005;**39**(8):1481-1488

[13] Cavinato C, Fatone F, Bolzonella D, Pavan P. Thermophilic anaerobic co-digestion of cattle manure with agro-wastes and energy crops: Comparison of pilot and full scale experiences. Bioresource Technology. 2010;**101**(2):545-550

[14] Di Maria F, Sordi A, Cirulli G, Gigliotti G, Massaccesi L, Cucina M. Co-treatment of fruit and vegetable waste in sludge digesters. An analysis of the relationship among bio-methane generation, process stability and digestate phytotoxicity. Waste Management. 2014;**34**(9):1603-1608

[15] Edwiges T, Frare LM, Alino JHL, Triolo JM, Flotats X, de Mendonça Costa MSS. Methane potential of fruit and vegetable waste: An evaluation of the semi-continuous anaerobic mono-digestion. Environmental Technology. 2018;**41**(7):921-930

[16] Ganesh R, Torrijos M, Sousbie P, Lugardon A, Steyer JP, Delgenes JP. Single-phase and two-phase anaerobic digestion of fruit and vegetable waste: Comparison of start-up, reactor stability and process performance. Waste Management. 2014;**34**(5):875-885

[17] Lindorfer H, Waltenberger R, Köllner K, Braun R, Kirchmayr R. New data on temperature optimum and temperature changes in energy crop digesters. Bioresource Technology. 2008;**99**(15):7011-7019

[18] Boe K, Batstone DJ, Steyer J-P, Angelidaki I. State indicators for monitoring the anaerobic digestion process. Water Research. 2010;**44**(20):5973-5980

[19] Fragoso R, Henriques AC, Ochando-Pulido J, Smozinski N, Duarte E. Enhanced biomethane production by co-digestion of mixed sewage sludge and dephenolised two-phase olive pomace. Waste Management & Research. 2021;**40**(5):565-574

[20] Schnürer A, Jarvis Å. Microbiology of the Biogas Process. Uppsala, Sweden: Swedish University of Agricultural Sciences; 2018

[21] Xu R-z, Fang S, Zhang L, Huang W, Shao Q, Fang F, et al. Distribution patterns of functional microbial community in anaerobic digesters under different operational circumstances: A review. Bioresource Technology. 2021;**341**:125823

[22] Mozhiarasi V. Overview of pretreatment technologies on vegetable, fruit and flower market wastes disintegration and bioenergy potential: Indian scenario. Chemosphere. 2022;**288**:132604

[23] Arhoun B, Villen-Guzman M, Gomez-Lahoz C, Rodriguez-Maroto JM, Garcia-Herruzo F, Vereda-Alonso C. Anaerobic co-digestion of mixed sewage sludge and fruits and vegetable wholesale market waste: Composition and seasonality effect. Journal of Water Process Engineering. 2019;**31**:100848

[24] Xu R, Zhang K, Liu P, Khan A, Xiong J, Tian F, et al. A critical review on the interaction of substrate nutrient balance and microbial community structure and function in anaerobic co-digestion. Bioresource Technology. 2018;**247**:1119-1127

[25] Martín-González L, Font X, Vicent T. Alkalinity ratios to identify process imbalances in anaerobic digesters treating source-sorted organic fraction of municipal wastes. Biochemical Engineering Journal. 2013;**76**:1-5

[26] Arhoun B, Villen-Guzman MD, Vereda-Alonso C, Rodriguez-Maroto JM, Garcia-Herruzo F, Gomez-Lahoz C. Anaerobic co-digestion of municipal sewage sludge and fruit/vegetable waste: Effect of different mixtures on digester stability and methane yield. Journal of Environmental Science and Health, Part A. 2019;**54**(7):628-634

[27] Ezieke AH, Serrano A, Clarke W, Villa-Gomez DK. Bottom ash from

smouldered digestate and coconut coir as an alkalinity supplement for the anaerobic digestion of fruit waste. Chemosphere. 2022;**296**:134049

[28] Casallas-Ojeda MR, Marmolejo-Rebellón LF, Torres-Lozada P. Identification of factors and variables that influence the anaerobic digestion of municipal biowaste and food waste. Waste and Biomass Valorization. 2021;**12**(6):2889-2904

[29] Astals S, Nolla-Ardèvol V, Mata-Alvarez J. Anaerobic co-digestion of pig manure and crude glycerol at mesophilic conditions: Biogas and digestate. Bioresource Technology. 2012;**110**:63-70

[30] Bouallagui H, Touhami Y, Cheikh RB, Hamdi M. Bioreactor performance in anaerobic digestion of fruit and vegetable wastes. Process Biochemistry. 2005;**40**(3-4):989-995

[31] Montañés R, Pérez M, Solera R. Anaerobic mesophilic co-digestion of sewage sludge and sugar beet pulp lixiviation in batch reactors: Effect of pH control. Chemical Engineering Journal. 2014;**255**:492-499

[32] Pullammanappallil PC, Chynoweth DP, Lyberatos G, Svoronos SA. Stable performance of anaerobic digestion in the presence of a high concentration of propionic acid. Bioresource Technology. 2001;**78**(2):165-169

[33] Wang Y, Zhang Y, Wang J, Meng L. Effects of volatile fatty acid concentrations on methane yield and methanogenic bacteria. Biomass and Bioenergy. 2009;**33**(5):848-853

[34] Scano EA, Asquer C, Pistis A, Ortu L, Demontis V, Cocco D. Biogas from anaerobic digestion of fruit and vegetable wastes: Experimental results on pilot-scale and preliminary performance evaluation of a full-scale power plant. Energy Conversion and Management. 2014;**77**:22-30

[35] Cabbai V, De Bortoli N, Goi D. Pilot plant experience on anaerobic codigestion of source selected OFMSW and sewage sludge. Waste Management. 2016;**49**:47-54

[36] Nguyen M-LT, Hung P-C, Vo T-P, Lay C-H, Lin C-Y. Effect of food to microorganisms (F/M) ratio on biohythane production via single-stage dark fermentation. International Journal of Hydrogen Energy. 2021;**46**(20):11313-11324

[37] Hadiyarto A, Budiyono B, Djohari S, Hutama I, Hasyim W. The effect of F/M ratio to the anaerobic decomposition of biogas production from fish offal waste. Waste Technology. 2015;**3**(2):58-61

[38] Azevedo A, Gominho J, Duarte E. Performance of anaerobic co-digestion of pig slurry with pineapple (ananas comosus) bio-waste residues. Waste and Biomass Valorization. 2021;**12**(1):303-311

[39] Holliger C, Alves M, Andrade D, Angelidaki I, Astals S, Baier U, et al. Towards a standardization of biomethane potential tests. Water Science and Technology. 2016;**74**(11):2515-2522

[40] Carvalho A, Fragoso R, Gominho J, Duarte E. Effect of minimizing d-limonene compound on anaerobic co-digestion feeding mixtures to improve methane yield. Waste and Biomass Valorization. 2019;**10**(1):75-83

[41] Alibardi L, Cossu R. Effects of carbohydrate, protein and lipid content of organic waste on hydrogen production and fermentation products. Waste Management. 2016;**47**:69-77

[42] Zahedi S, Sales D, Romero LI, Solera R. Optimisation of the two-phase dry-thermophilic anaerobic digestion process of sulphate-containing municipal solid waste: Population dynamics. Bioresource Technology. 2013;**148**:443-452

[43] Cheng H, Li Y, Guo G, Zhang T, Qin Y, Hao T, et al. Advanced methanogenic performance and fouling mechanism investigation of a high-solid anaerobic membrane bioreactor (AnMBR) for the co-digestion of food waste and sewage sludge. Water Research. 2020;**187**:116436

[44] Montero B, García-Morales JL, Sales D, Solera R. Analysis of methanogenic activity in a thermophilic-dry anaerobic reactor: Use of fluorescent in situ hybridization. Waste Management. 2009;**29**(3):1144-1151

[45] Carvalheira M, Cassidy J, Ribeiro JM, Oliveira BA, Freitas EB, Roca C, et al. Performance of a two-stage anaerobic digestion system treating fruit pulp waste: The impact of substrate shift and operational conditions. Waste Management. 2018;**78**:434-445

[46] Elsayed M, Andres Y, Blel W. Anaerobic co-digestion of linen, sugar beet pulp, and wheat straw with cow manure: Effects of mixing ratio and transient change of co-substrate. Biomass Conversion and Biorefinery. 2022:1-10

[47] Mateus S, Carvalheira M, Cassidy J, Freitas E, Oehmen A, Reis MAM. Two-stage anaerobic digestion system treating different seasonal fruit pulp wastes: Impact on biogas and hydrogen production and total energy recovery potential. Biomass and Bioenergy. 2020;**141**:105694

[48] Fonoll X, Astals S, Dosta J, Mata-Alvarez J. Impact of paper and cardboard suppression on OFMSW anaerobic digestion. Waste Management. 2016;**56**:100-105

[49] Garcia-Pena EI, Parameswaran P, Kang DW, Canul-Chan M, Krajmalnik-Brown R. Anaerobic digestion and co-digestion processes of vegetable and fruit residues: Process and microbial ecology. Bioresource Technology. 2011;**102**(20):9447-9455

Chapter 8

Waste Separation and 3R's Principles for Sustainable SWM: Practice of Households, Private Companies and Municipalities in Five Ethiopian Cities

Abdulkerim Ahmed and Meine Pieter van Dijk

Abstract

In many developing countries, solid waste is not collected and disposed properly. This leads to public health risks; for example, due to water contamination and air pollution. This research examines how municipalities frame and act on waste separation and 3'R principles (reduce, reuse and recycling) and how they involve households and private providers to ensure sustainable SWM. Questionnaire, interview and Focus Group Discussions (FGD) were used to gather data from private and government municipal service providers. The findings indicated that, in Ethiopia, the federal rules, policies and contractual arrangements for SWC clearly show the necessity for waste separation and 3'R principles. Yet, the lack of commitment of the Cleaning Administration Agency (CAA) and Cleaning Administration Department (CAD), the less concern of the private companies and the residents toward the absence of responsible institution to integrate actors are existing challenges. The Ethiopian government needs to improve the governing capacity of municipalities. Besides, it should commit a full-scale implementation of sustainable SWM through raising public awareness of waste separation, the 3R's principles, along with polluters' pay principles. Otherwise, unnecessary loss of resources and overwhelming adverse impact on the environment, Ethiopian public health and safety will persist prevailing.

Keywords: environment, awareness, waste separation, reduction, reuse, recycling

1. Introduction

Henry and Yongsheng [1] state the primary target of municipal SWM is to protect the health of the population, promote environmental quality, develop sustainability and provide support to economic productivity.

As with many Sub-Saharan African cities [2–4], Ethiopia has experienced many problems related to SWM for a long time. Solid waste accumulation within cities raises concerns about public health, environmental pollution and detrimental fatalities. The uncollected solid waste turns to be an ideal place for breeding microbial pathogens, flies, rats and other creatures which spread diseases [5], such as cholera, diarrhoea, etc. Poor hygienic practices have contributed to the spread of diseases in Ethiopia which accounts for more than 60% of the disease burden in the country [6]. The recent accidental death on 'Koshe' disposal site is evident for the destructive nature of unmanaged waste.

Municipal services in developing countries are handicapped by limited finances. The ever-increasing demand on urban services has also exerted another challenge. As a result, contracting out SWC services to the private sector has emerged to fill the gap in service delivery.

The gathering and safe removal of waste require due emphasis. For example, generation rates in 2009 in Nairobi are estimated at 1850 tons/day. Only 33% is composed and removed [7]. Similarly, according to Okot-Okum, only 21.7% is gathered in a daily base. Likewise, from 9,240,000 kg waste collected daily, only 29% is gathered in Ethiopia. This shows that various areas of the cities do not obtain appropriate service from SWC. As a result, most citizens in the urban areas dispose their waste randomly. This also indicates that the waste separation and 3 R's principles should be properly conceived and implemented. In developing countries, waste is not processed, reused in a cost-efficient and safe manner [8].

There are a number of academic studies on SWM in Africa. To mention some, Awortwi [9] focuses on governance in multiple arrangements and the relationships between capacity and contractual arrangements. Obirih-Opake [4] deals with on public private partnerships (PPP), specifically the impact of decentralisation and private sector participation on urban environmental management. The concerns for Karanja [8] are sustainable development issues in SWM; institutional arrangements; the role, interest, success and failure of different actors. Collins [10] and Koppenjan [11] show the challenges facing PPP and identify the focus of private sector as only on short-term return on investment. Katusiimeh [12] indicates the lack of regulation and transparency in Uganda. Tilaye and van Dijk [13] underline the importance of consistent power of the state in shaping the developmental role of the private sector (focus on micro enterprises). Baud and Post try to connect SWM with sustainable development by operationalising three broad goals: ecological sustainability, socio-economic equality and improvement of health. They argue that there is a gap in the current literature on sustainable SWM in developing countries that the system is rarely investigated in its entirety. Assessments combining separation of waste, productive use of waste and the three broad goals of sustainability are still largely absent [14]. The number of studies on SWM remark the importance of regulatory and institutional framework, and they have shown the rhetoric nature of the issues when compared with the real practice on the ground. The sustainability of SWM depends on how the waste separation and the 3R's principles are conceived and implemented by actors in the waste management sector. This research examines how municipalities frame and act on waste separation and 3'R principles and how they involve households and private providers to ensure sustainable SWM. The rational to examine this issue is due to the fact that the sustainability of SWM at the local government level is greatly influenced by the role the aforementioned actors play.

This study hypothesised that the current SWC practice through PSI fails to assimilate waste separation and the 3R's principles and make use of the role the actors could play for sustainable SWM. Thus, the research question is: How have the local governments framed and acted on waste separation and the 3R's principles and engaged the actors for sustainable SWM? We offer insights into the ways of making waste separation and 3'R s principles feasible through the integrated involvement of household and private providers.

2. Theories on sustainable SWM, waste separation and the 3 R's principles

The purpose of sustainability in SWC is to give good services for humans and protecting the environments [15]. It is a theory regarding the stability of economic, social and environmental features. According to Van de Klundert and Anschütz [16], sustainable SWM contains a system that can keep itself for longer periods without decreasing the resources.

In defining sustainable SWM, financial, social and environmental issues are essential elements.

The idea involves that environmental sustainability needs SWC and disposal that impose an unlimited load on the environment to be considered as resources which could be transformed in a closed-cycle system by restoring numerous natural cycles that in turn leads to a smaller loss of raw materials, energy and nutrients.

The application of waste separation and the 3R's principles in SWM practices depends on some considerations. Waste reduction is associated with the knowledge of waste generators reuse and recycling. Essentially, recycling behaviours have a direct link with three sets of variables. Firstly, according to Vining and Ebreo [17], with environmental values form underlying attitudes towards the environment. In such attitude, people are more likely to act in environmentally suitable behaviours. Secondly, with situational aspects which hold facilitating and spiking impacts and are usually categorised as contextual [18, 19], socio-demographic [20], knowledge-centred [21] and experience-centred [22]. Thirdly, with psychological aspects that are distinctive perceptual characters of the individual and contain altruistic drives to reuse [23], outcome beliefs [23], subjective norms (or social influence; Chan [24]). Other related issues are a set of logistical factors and personal efficacy variables, such as time to undertake the activity [25] and belief that individuals have a responsibility to protect their environment which is termed as environmental 'citizenship' beliefs [26].

The core significance of recycling, composting and other commodity-based actions is the decrease in the quantity that wants to be moved to dump areas. In various countries, still there is a vast difference between the reality in the ground and the principle. From an environmental viewpoint, recycling may be favoured, but the economic costs or the problems of the institutional may inhibit waste recycling implementation in integrated SWM. The real integration can occur at numerous stages, Van der Klundert and Lardinois [27]. This contains waste processors such as formal and informal recyclers; waste generators such as households, industry and agriculture; and government organisations such as waste managers and urban planners. By collaborating with the societies, micro-enterprises may play a great role in public environment education. Having strong relation with the inhabitants can offer chances to introduce separation at sources. This could help workers benefit from collecting and selling the recyclable materials and create an opportunity for cost reduction for the municipality by reducing the quantity of waste.

3. The conceptual framework

The conceptual framework of sustainable SWM above is designed by the author after reviewing the theoretical perspectives raised so far. The cornerstones of the framework are environmental awareness of actors, institutional and regulatory arrangement. The logical process within the framework flaws from environmental

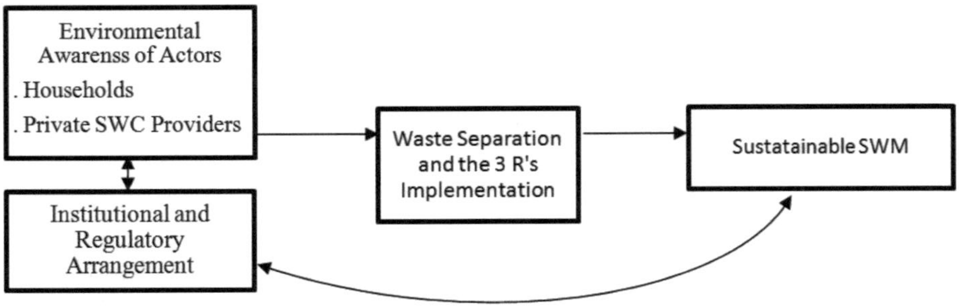

Figure 1.
Conceptual framework for implementing sustainable solid waste collection through 3R's.

awareness to organisational arrangement for the implementation of bylaws. The basis of the framework hinges on the notion of a link of factors, environmental awareness, institutional arrangement [28]. Thus, a flexible but a structured framework is formed to conceptualise previous studies in the area and also to organise data collection and analysis.

We consider four components (C1–C4, see **Figure 1**) for analysis. They are the environmental awareness of households and private SWC service providers for waste separation and the 3R's (C1), the institutional, regulatory arrangement and availability of infrastructure, rules for the 3R's, contractual obligations between CAA and CADs regarding 3R's; moreover, CAA's and CADs' enforcement capacities are examined [29]. This component also considers the integration among stakeholders (C2). Then, (C3) looks at the implementation of waste separation and 3R's. (C4) evaluates the outcomes of the SWM sector in terms of sustainability. In this component, households and private providers' awareness and the integration of such actors and the contents in the contract are evaluated whether they lead to the implementation of waste separation and the 3R's principles and environmental sustainability. Analysing the components, we provide recommendations for the feasibility of waste separation and the 3R's principles implementation which lead to achieving sustainable SWM in Ethiopian cities. This could be relevant for other developing countries as well.

4. Description of the study area

The 2003 Constitution of Ethiopia stipulates that the Federal Democratic Republic of Ethiopia (FDRE) comprises nine administrative regions and two administrative council cities. The total population of the country is about 90,078,000 [30]. There are 145 cities and towns with a population of more than 15,000 inhabitants. The rapid, constant growth of the urban population has led to a striking increase in urban solid waste generation, with a crucial environmental and socio-economic (i.e. related to disease and wastage of resources) impact. Low levels of access to SWC facilities and poor hygienic practices have contributed to the spread of diseases in Ethiopia. These factors account for more than 60% of the disease burden in the country [6]. Five cities were chosen for this study: namely Addis Ababa, Bahir Dar, Mekelle, Hawasa and Adama. These cities face rapid urbanisation and are the main regional capitals with the largest urban populations endangering the environment by generating much solid waste.

A previous study in Addis Ababa by Debere et al. [31] found that the median waste generation rate varied per household from 0.361 to 0.669 kg/day. Tadesse et al. [32] found a variation between 0.30 kg/day and 0.33 kg/day/household in Mekelle. The review of a baseline survey report shows that the average household in Hawasa city generates 0.42 kg/day. Getahun and Mengistie [33] indicates that low, middle and high-income households in Adama generate 0.67 kg/day, 1.21 kg/day and 1.87 kg/day, respectively. The per capita waste generation in Bahir Dar city is 0.25 kg/day from residential sources; however, it is about 0.45 kg/day from combined municipal waste – from residential, commercial, institutional and street sweeping sources [34]. The solid waste generation rate in Ethiopian cities is similar to that of low-income countries. Glawe et al. [35, 36] confirm that the waste generation rate of low-income countries is 0.1–0.5 kg/capita/day as opposed to 1.1 kg/capita/day and above in fully industrialised countries.

5. Materials and methods

Survey questionnaires, interviews, focus group discussion, desktop survey and observation were used in this study. We collected data during two fieldwork periods in five Ethiopian cites: Addis Ababa, Mekelle, Bahir Dar, Hawas and Adama, between June 2014 and March 2017.

The researchers used a pre-tested structured questionnaire to gather information from 40 companies. Half of the 40 registered private companies were taken from Addis Ababa CAA, and the other three CADs were from the municipalities of the remaining four cities. The questions in the questionnaire were measured on a five-point Likert scale with categories ranging from 'Strongly disagree', 'Disagree', 'Cannot tell', 'Agree' and 'Strongly agree', 'Very poor = 1,"Poor = 2', 'Fair = 3', 'Good = 4 'Very good = 5'. Also, Binary Scale (Yes/No) questions were included [37]. Analyses were conducted on secondary sources comprising available archive records and document from private companies' monthly reports, contract agreement documents and CAA and CADs official waste-related documents.

Our interviewees were government experts from Ethiopian Environmental Protection Agency Ministry of Environment and Forest (EPA), Addis Ababa CAA and EPESA. We also visited company managers who gave us detailed information regarding their organisation's SWC practice. It is believed that managers initiate, devise and oversee the operations of their company. The interview guide for government officials was open-ended questions and Yes/No questions.

In this study, Focused Group Discussions (FGDs) were conducted the five cities. The responses remained open-ended, and this enabled respondents to produce as much information as they wish to provide. This approach allows respondents to describe and analyse their experiences or feelings in their own words without being constrained by any form of framework.

Two FGD groups were formed. The first group had a total of six individuals with a composition of participants of whom two urban health extension workers, two health development representatives army – one male and one female. Community representatives were one male and one female. This group represented service users. The other FGD group consisted of six participants – two municipal officials, two private companies and two MSE members. This group represented service providers. Generally, the two groups were formed so as to make them free to respond genuinely. Members of the first group were taken through the communication with kebelle officials since they work with the kebelle representing residents. Both the two FGDs lasted

80 minutes each in average. Moreover, for the purpose of triangulation 25 households, five in each city were interviewed. These households were selected after proving that they had lived in each city more than 5 years. The justification was based on the belief the more they live in each city, the more they know about the SWC practice.

The observation was made by visiting throughout the city both in planned and unplanned areas and by taking a view of the whole process of SWC and general cleanliness. This gave us an opportunity to observe the practices within households and private providers in the institutions, transfer stations and disposal sites.

A combination of both qualitative and quantitative methods provided the most accurate data for this study. Data were categorised and cross-tabulated. To get accurate description of SWC practices, official documents and interview results were analysed qualitatively using narration. The methodological part of this paper did not use mathematical models for analysing the data which could have strengthened the result of the study, this may be stare as a curb for this study, but this can be improved by the next research studies on the topic.

6. Findings and discussion

6.1 SWC practices in the five cities

Solid waste is kept in waste sacks (locally knows as 'kesha') and containers 'Genda' before collection. Most of the waste sacks are over-used, with some holes on them uncovered and untidy. As a result, leachate and odour are unpleasantly common. The situation is aggravated by the presence of ownerless dogs that disperse the collected waste at the transfer station while searching for food. It was also observed that the wind disperses the dirt. Only a few households in each city use standard containers, durable with fitted lids, able to prevent odour and leachate flow from the waste, and these are mainly in high-income areas.

It is common in the five cities to see wastes in open drains or on the ground near the houses before the collection truck takes them. The waste is collected by the crew members and transferred to the disposal sites. These workers often complained about the littering habit of households, i.e. pushing waste from their houses into the ground, drainage lines and on roads. Burning and burying solid waste around home areas are practised by some households. During collection, crew members move around the households to alert them with alarms to bring waste out for collection. Then, they unload the waste from the waste sacks and containers, put them in the collection trucks and return the waste sacks to the owners. House-to-house and communal type of SWC is practised in the study cities. Small and Medium Enterprises (SMEs) in Addis Ababa and in Adama collect waste from households, whereas SMEs in Bahir Dar and Hawasa collect waste from households, hotels and institutions and take the waste to transfer stations. In Addis Ababa, waste from hotels and institutions is taken to landfills by private companies while such waste is taken to transfer stations and landfills by companies in the other cities. CAA's and CADs' vehicles take waste from transfer points (communal containers) to the disposal site.

6.2 Awareness

FGD participants revealed that several individuals put waste in waste sacks and throw it in the narrow streets or under the fence of other households. Despite the presence of dust bins in most central parts of the city either as standing on the road or mounted on lamp posts or telegraph poles and other strategic sites, such as bus stops, shopping places and walk streets, people do not use them properly. As a result, banana skins, bus tickets, used tissue paper, chewing gums, etc. are seen on the street. And individuals are not held responsible for such trashes.

FGD participants especially in Addis Ababa confirmed that travellers and strangers are not concerned about the cleanliness of the city since they lack sense of ownership and awareness. From the FGD, we noticed that most of the time it is the women who manage and throw out the waste. Culturally, SWC tasks are largely left to mothers and maid servants. However, Hage et al. [38] found that gender does not matter much for recycling behaviour. Most households fail to recognise channelling of materials at household level to SMEs as precursor for supplying of material input to recycling and reuse process. We also noticed that people have low perceptions of the waste problem and its threat and their responsibility of safe environment.

Interviewed private company managers and SMEs members complained about health and safety issues due to poor SWC system. They also showed dissatisfaction with the poor cooperation of residents and poor storage practice of some households.

A study by Tilye and van dijk [13], which is related to this, revealed that the city cleanliness index of Addis Ababa was much less than that of the neighbourhood index. This was due to lack of public awareness of the health implications of unsanitary practices and residents' indifference to the presence of waste.

FGD results indicate that CAA and CADs neither have they done much on public education themselves nor have integrated with service providers to do so. CAA and CADs did not provide the residents with facilities, such as dust bins and containers for sorting waste. Moreover, their failure to enforce a regulatory sanction made residents become less careful and even ignorant for waste handling. More general education and information about the broader issues underlying waste management are required in order to explain to the public the need for and the benefits from the acceptance of a wider responsibility towards waste disposal [39].

6.3 Awareness for waste reduction

What is ideal is reducing waste generation beforehand so that no waste is left for disposal or recycle. However, this conception is not equally understood by the community. In all the five cities, FGD participants mentioned that most of the residents are unaware of waste reduction. For instance, households are unaware of the importance of avoiding small individual packages of any products and using of durable plastic materials for handling as part of waste reduction activities. Even though the Ethiopian saying 'Alemakosheshmatsdatnaw/not generating waste is tantamount to keeping clean'is displayed in public places, but it seems that most people do not value it.

Failing to understand the benefit of waste reduction is still seen even among the SW collecting private companies and SMEs. The main reason for this is, according to the FGD, their payment is based on the amount of waste they collect and dispose. Some companies even believe that the application of waste reduction means an activity which hinders their business as, for instance, compost making reduces the amount of waste to be collected. Even SMEs and other companies do not want residents to bury or incinerate waste. Some companies even mix ashes, concrete scarps with other waste so that the waste becomes heavier and looks full.'

6.4 Practice of waste separation

Ecological sustainability concerns household participation in reuse, recycling and composting and the intention to support recycling by practising in separation at source. The FGD results and personal observations show that many households do not have awareness about waste separation. An interviewee from a household in Mekelle said, 'Let alone waste separation our district has no coverage for SWC', he added, 'we are better than others because we collect waste at least in a waste sack and make it ready for collection'. A few households reasoned lack of sorting materials for their failure of separating waste.

A household interviewee in Addis Ababa said people in the area do not have a warranty to get back their own relatively new waste storing material from SMEs. This forced them to use less value materials which subsequently affect the quality of waste handling. Two household interviewees who live in a condominium in Adama said they do not want more than one waste sack reasoning lack of space in their compounds.

Households are required to have two solid waste collection receptacles – one for organic and the other for non-organic waste as gathered from the FGD in Mekelle. However, this was later stopped because of residents' carelessness and absence of enforcement from CADs.

As a better practice, in a sub-district of Bahir Dar city, people used to separate waste using three different coloured containers, i.e. yellow for plastic waste; green for degradable waste; and black for hazard waste. Officials in the city confirmed that there was an effort by individuals at household level to separate waste for different purposes. There were trained households by government, civil society and community-based organisation to separate the organic part of the waste and produce compost in their compounds. This was later stopped because of the residents' carelessness and absence of support from CADs. Some households separate animal extract and use or sell it for fuel purpose.

According to the CAA and CADs officials, the challenges for waste separation are CAA's incapability to provide households with different coloured waste sorting materials even in the rich neighbourhoods and the absence of a responsible monitoring body. All the cities shared the lack of kerbside recycling bins, lack of recycling machineries both in the transfer stations and in the disposal sites and a gap towards mobilising household and private providers to separate waste. This finding is similar to Williams and Kelly [40], who found that people mentioning insufficient availability of space to store recyclables both inside and outside the home as well as inadequate local facilities as barriers for separation and recycling. However, CAD officials in Mekelle planned to request their regional administration to produce plastic waste bins to be distributed to the residents at a lower price.

6.5 Waste reuse and recycling practice

It was noticed from the FGD that willingness to reuse and eventually reduce waste is not based on the positive environmental attitudes and active concern for waste issues; it is rather on economic benefits. This is because they use the waste to earn money. Households are the first level for identifying the type of waste for reuse. This practice even extends to the disposal sites. Disposal sites do not have fences. People live around them. Painstakingly, some people search for materials with their naked fingers to extract plastic bags, cans and iron bars.

Recyclable materials in the household waste stream in each city can be categorised as dry, i.e. comprising paper, plastics, metals, glass and textile, etc. and organic material, consisting of kitchen (food) and garden waste, which are suitable for composting. From our observation in the five cities, we noticed private companies were collecting solid waste with high organic content.

A resident interviewee in the rich neighbourhood said that there is potentially recyclable waste in his compound; for instance, shampoo bottles, empty facial tissue boxes, empty toilet paper tubes and the like. Items that have intrinsic commodity value, such as glasses, cans, rubbers, plastics, low grade paper, are locally traded. These materials are bought by itinerant buyers (Locally known as 'korale' and 'Liwach'). The 'korale' are door-to-door traders who buy and/or exchange reusable items; for example, tins, plastics, bottles, nail varnish containers, broken cooking jars, used shoes, old garments, etc. They supply such materials to middlemen at Mercato in Addis Ababa and at the local markets in the other cities.

These traders have a huge impact on the reduction of solid waste both at household and city level. Despite their well-established economic system and a market niche in the city, they were less recognised by the public sector for their contribution to recycling and reuse. According to the FGD with CAA and CAD officials, such individuals are informal collectors from the poor section of the society. There are more than 10,000 informal collectors in Addis Ababa as secondary data indicated. Yet, they are not organised and do not have support from the city administration.

A review of the 'Urban Development and Construction Minster, Waste Management and Green Development Strategy Document' shows that Ethiopia's climate Resilient Green Economy strategy underpins mainly four issues: forest protection and development; electric power generation from renewable energy; modern energy saving transport, and industry development. Linked to waste management, the strategy stipulates wastes from households, transfer stations and disposal sites should be reused to enhance agricultural productivity through compost and bio gas production. A CAA official in Addis Ababa mentioned that in the disposal site called 'Koshe', biogas generation project is undergoing.

This strategy document states that SME associations who are involved in the reuse and compost production could be granted land with discounted lease payment and facilitated. For instance, capital goods imported for such production will be tax-free. This makes SMEs and households to be integrated into compost marketing. So, the benefit goes both to them and to the farmers through satisfying their demand for fertilisers. Nevertheless, the FGD shows that the strategy has not yet been successfully implemented. SME interviewees confirmed that though they were conscious about the strategy, they were not involved and became beneficiaries. On the other hand, some households consider the compost production as a burden on their livelihood, and they fear they may not have a market for it. Still, some others complained about a lack of space to produce compost.

As a good practice in the periphery of Bahir Dar city, we discovered a resident who locally produces compost from organic materials consisting of kitchen and garden waste. He uses his compost in his garden and farmland. He also sells the compost in the local market. He said he had a vision of environmental responsibility. He also envisioned that he would produce more if he would get machinery support as he believes composting is a business opportunity. He remarked that if people are educated in this regard, over the years people gradually would evolve from a hauler of waste to a producer of fertilisers.

Connected to reuse and recycling, a large proportion of reusable materials are transported to small towns and rural areas to be used as household items, such as plastic bottles for holding water, oil and gas; used cans for storing cereals, honey, butter and spices. In the five cities, there are industries that reuse rejected paper, glasses, plastic fabrics, iron pieces and steel rods as raw materials. Raw materials for plastic fabrics which are imported could have been substituted by local scraps through recycling.

7. Integrationamong actors for waste separation and 3R's principle's

The partnership mainly among municipalities, private providers and households is essential for sustainable SWM. However, the FGD revealed that there is a missing link among these actors in all the study cities. Out of 40 companies surveyed, 32 admitted that they have never had meetings with households, whereas eight companies said they used to have a few meetings with households.

When interviewed, CAA and CADs officials in the study cities contended that they advised households via the local governments (Kebele) to use micro-enterprises for primary SWC and carry out waste separation. However, they admitted that they did not schedule for discussion and communication between the community and the SMEs as well as private companies. It is the government which pays and monitors the SWC system. Households do not own the process. Moreover, health extension workers participated in the FGD mentioned that there has not been integration among government sector offices such as health office, municipalities and private providers. It was also discussed that CAA and CADs are supposed to ensure co-operation among the actors to implement waste separation and the 3R's principle's.

For the FGD participants, the design and operation of SWC are managed in a top-down approach. There is a by-law for waste separation and 3R'sat CAA and CADs level. However, the SWC system including the service payment for private providers is not subject to this by-law. The role of actors to implement the by-law is overlooked. Consequently, households do have less participation in the scheme. The lack of awareness is the major contributory factor for the failure of most SWM schemes [41]. Post and Obirih-Opareh [42] report that to make SWC services more sustainable, consulting residents about the most appropriate methods of collection, i.e. affordability of materials for waste separation and cost sharing is important.

8. Companies' views on waste separation and 3R's practice

Companies were asked whether they separate waste and implement the 3R's principle's, the views of the company managers are presented in the **Table 1**.

As shown in the table above, 31 of the private companies disagree that they separate waste; eight companies said they used to separate. A company manager in Addis Ababa said they stopped separating because of the difficulties to provide different waste sacks (plastics) to households and the unwillingness of households to separate waste. Another company manger in Mekelle said as they have limited number of waste collecting vehicles and crew members, hence it is difficult to arrange and manage different vehicles for separated waste. He said that is why they stopped early even though their agreement with the CAD needs them to separate waste.

Indicators	Agree					Disagree				
	Addis Ababa	Mekelle	Adama	Bahir Dar	Hawasa	Addis Ababa	Mekelle	Adama	Bahir Dar	Hawasa
Do you separate waste	4	2	1	1	1	16	2	5	5	4
Do you have meetings with customers to discuss about SWC	2	2	2	1	1	18	2	4	4	4
Conducts public education regularly	20	2	2	2	3	0	2	4	3	2
Does the contract include 3R principles										

Table 1.
Companies' views on waste separation and 3R's practice.

Company manger complained about the attitude of the people towards waste separation. Interview results show that some people have even negative attitudes towards waste collection.

One company manager told us that some people are careless for collecting waste let alone sorting it. Another one added that doors are often found closed during waste collection time.

9. Public education for sustainable SWM

CAA and CADs are obliged to conduct public education regularly on sound environmental practices. Although 29 out of the 40 companies agree with the presence of public education, they said that it is not implemented regularly. From the FGD we noticed that, in Ethiopian cities there is no public education about 3R principles even on TV or in the school's curriculum. Changing individual behaviour towards recycling and other forms of waste management is central to achieving a sustainable future [43, 44]. That is why such effective programmes have to be implemented to match this goal.

There are fundamental principles of the by-laws as the document review of Environmental Protection Agency (EPA discloses. Accordingly, principle 'A' states: Cities have the relevant, responsibility to ensure that all waste generated within their jurisdiction is collected, transported treated, disposed of or recycled. And principle 'B' underlines that such collection, transportation, treatment, disposal or recycling is the responsibility of waste management hierarchy as stated in the subsections of the by-laws.

The by-laws underpin principles to establish a waste management hierarchy of priority: (a) waste avoidance, waste minimisation and waste reduction; (b) reuse; (c) recycling, reprocessing and treatment; and (d) disposal. However, the principles are not properly implemented by CAA and CADs as the officials confirmed during the interview. For example, there is no sanctioning against offenders, against residents and private providers for violating the by-laws. Companies' views on CAA and CADs' implementation based on their contractual agreement are presented.

Monitoring of SWC needs information about performance targets type of waste collected, number of trips made and cubic meter of waste disposed.

As it is shown in **Table 2**, 32 respondents said monitoring of the companies' compliance to regulations and standards on 3R principles was inadequate. Quantity of waste collected and their output by CAA and CADs was adequate according to 36 companies' responses. They have workers appointed in the disposal sites to keep records on the waste disposed. The payment is based on these records. However, there is no effort from CAA and CADs to monitor companies to separate waste.

The CAA and CADs have an obligation of supervising the private companies to implement 3R principles. The contract management and supervision of the private sector by the CAA and CADs are rated inadequate by 37 companies. The interview with these companies shows this is due to the limited capacity and lack of commitment of CAA and CADs to enforce the principles.

Monitoring of SWC includes daily inspection of service areas and container sites; issuing warning letters and terminating contracts for poor performance.

CAA and CADs do not adequately enforce the sanctions indicated in the contract as 34 of the respondents pointed out. Oduro-Kwarting [37] remarks the experience from Ghana shows the municipality takes responsibility for monitoring the performance of service providers. Providers shall co-operate fully with the monitoring firm. This means, it allows the firm to have access at all times; to inspect work being carried out under the agreement and all records and documents maintained by the service provider. It also permits the firm to inspect the service provider‘s vehicles, plant, stocks of spare parts and workshop facilities. This, in turn, helps to observe regulations. Moreover, providers shall attend monthly meetings at the request of the regulator firm where operational and other issues of mutual interest may be discussed.

CAA and CADs apply enforcement of by-laws and contract rules on companies. On this issue, the views of the company managers are presented in **Table 3**. They do have differing views as 34 of them disagree the by-laws on 3R were enforced; 37 of them also disagree that the fines for defaulting the by-laws were not punitive enough; 38 of them assert that the monitoring of compliance was not effectively done, and 35 of the companies said the environmental health standards and sanitary regulations implementation of 3R were not strictly observed and enforced.

10. Challenges and opportunities for waste separation and 3R's implementation

An interview with an official in the EPA disclosed that even though the underlying principles to establish a waste management hierarchy in the order of priority of importance: (a) waste avoidance, waste minimisation and waste reduction; (b) reuse; (c) recycling, reprocessing and treatment; and (d) disposal is already documented nationally. Yet, the implementation of these principles has not been feasible due to constraints, mainly a failure to execute strict rules, such as fining residents for not separating waste. In fact, this issue is politically sensitive according to the informant. The way out as he suggested is educating the public. It is being planned to include sustainable SWM in school curriculum and other sectors' programmes. However, the FGD with service users shows that the local administrations are not committed. The capacity of the CAA and CADs to

Regulation indicators	Adequate					Inadequate				
	Addis Ababa	Mekelle	Adama	Bahir Dar	Hawasa	Addis Ababa	Mekelle	Adama	Bahir Dar	Hawasa
Monitoring of companies service quality and compliance to regulations and standards set in the contracts regarding (implementation of 3R)	2	4	2	0	0	18	0	4	0	0
Monitoring of households whether separating waste or not	20	4	0	5	5	0	0	6	0	0
Supervision of the private providers to implement 3R and waste separation	1	1	1	0	0	19	3	5	5	5
CAA and CADs' enforce sanctions in the contracts signed with the Companies regarding waste separation	2	4	0	0	0	18	0	6	5	5

Filed survey result, 2016.

Table 2.
Companies' views on regulation of quality of service and 3R implementation.

Indicators	Agree					Disagree					Can not tell
	Addis Ababa	Mekelle	Adama	Bahir Dar	Hawasa	Addis Ababa	Mekelle	Adama	Bahir Dar	Hawasa	
The by-laws of the CAA and CADs and national laws on environmental sanitation specially 3R and waste separation are enforced.	3	0	3	0	0	15	4	3	5	5	2
The fine for defaulting by-laws is punitive enough	0	0	3	0	0	20	4	3	5	5	
The monitoring of compliance to solid waste service standards set in the contracts is done effectively.	0	2	**0**	0	0	20	2	6	5	5	
The environmental health standards and sanitary regulations are strictly observed and enforced	2	1	2	0	0	18	3	4	5	5	

Table 3.
Companies' views on enforcement of legislation.

facilitate, regulate and monitor 3R's principles implementation is hampered by a number of factors. According to one of the CAA official, 'The main constraints for implementing waste separation and 3R's are weak capacity of CAA and lack of political will as officials are busy with other urgent routines'. Not only is there less concern for SWM, but also setting inadequate budget is one of the main challenge. In this regard, the survey shows that there is very low service payment for SWC providers in Ethiopian cities. The FGD with CAA officials in Addis Ababa unfolds CAA collects more than 30 million ETB or 1,421,800 USD per month through water bill from the city residents. Other FGD results have also shown that all cities, but Adama, collect service charge in the same way.

No matter how much the service payment in each city is, all companies and SMEs complain about the under payment. Interviewed CAA officials in Addis Ababa said that only 2 million ETB from the total of 30 million ETB collected is allocated for SWC. This implies that the rest 28 million ETB is not used for SWC. This is true in all the other cities.

The interview with SMEs' members indicated that the monthly income members get is unsatisfactory even to afford medical and education expenses. In connection to this, Wilson et al. [45] confirm that low salary is paid typically between US $25 and 50 per month for primary waste collectors. This means that SWC is given less priority by local governments in developing countries. From the interview with SMEs' associations and private companies, we noted that there was low service payment. This practice discourages private providers from becoming vigorous towards waste separation and 3R's and even not pursue in SWC business. The findings in the FGD indicate that despite government's plan concerning achievement of food security through forming SMEs' associations and involving them into waste collection, the practice in each city shows the needy people who work in waste management are not beneficiaries of social safety net.

As opposed to the challenges discussed so far, the FGD uncovers some opportunities to carry out waste separation and 3R's. To begin with, uniquely in Ethiopian cities, SMEs are already organised in the form of association to readily embark on the implementation governed by the existing by-laws. Secondly, the growing industries to be fed by recyclable materials have created potential market. Thirdly, the recent national horrendous waste-slide event at 'Koshe' disposal site which was a cause for the death of dozens of poor people could serve as a wakeup call for the nation as far as 3R's and SWM are concerned. Fourthly, the Kaizen strategy which is being preached and practised presently to systematically remove recyclable office waste, such as used papers, ink cartridges, packing canvases, irrecoverable machines, broken furniture in government organisations is likely to be explored in waste separation and 3R's.

11. Conclusions

The study evaluated the practice of waste separation and 3R's which are critical for attaining sustainable SWM. The challenges for implementing these principles on the part of households, private providers and local administrations perspective are identified.

The research upheld the theoretical base of sustainable SWM. As shown in the literature, a waste management system needs to be tuned to the local conditions and feasible from a technical, environmental, social, economic, financial, institutional and political perspective [46]. It should be able to maintain itself over time without exhausting the resources upon which it depends. This is highly dependent on the preparation of an enabling environment and implementation of waste separation and

3R's principles. The finding shows that waste minimisation is not really adhered due to lack of knowledge at the household level. More importantly, the households, private solid waste collectors and informal waste collectors are not mobilised towards waste separation and 3R's implementation. The current SWC contractual obligation fails to enforce waste separation and 3R's principles and achieve sustainable SWM. However, economic incentives play a more important role than ecological considerations in separating and channelling reusable and recyclable items from the waste stream at household level. Informal waste pickers and households selling recyclable waste are driven by their need of earning money. These actors supply materials for reprocessing with the absence of legitimate strategy and regulations. The recyclable rejects could have been eventually alight into usable products and fertilisers for which there is unending demand. This has made the market for recyclable materials insufficient and the recycling loop unclosed. Thus, mostly it remains in the dark.

There is no integration for implementing waste separation and 3R's principles among actors. In each city, there is a practice of waste muddling which affects environmental sustainability. There are gaps to apply waste prevention and reduction principles through utilising composition and suitability of the waste for certain types of treatments. Consequently, against the popular sustainability concept of people, planet and profit (as shown in the conceptual framework), the waste in the disposal site in each city forms a large pile. This poses health risks to the society that ultimately could affect the air and water supply on planet sustainability. Besides, the existing SWC is not profitable financially for it disregards the potential prevention of pollution by reduction of waste and other reasons. Such phenomenon is stated by [47] as it is poorly committed to prolong the lifespan of landfills [48]; and nor does it help to succeed at bringing down the cost of waste collection, transporting, recycling and disposal [49].

Local governments lack of committed account for the failure to implement waste separation and 3 R's. Policies aimed at encouraging waste reduction, reuse and recycling are not effectively implemented. This is consistent with our hypothesis that the current SWC practice through PSI fails to assimilate waste separation and the 3R's principles and make use of the role the actors could play for sustainable SWM. Also, it coincides with the theoretical implication of the absence of waste separation, implementation if 3R's principles and the missing of the productive use of waste damage ecological sustainability, socio-economic equality and improvement of health.

The large heaps of solid waste left uncollected in Ethiopian cities are still dangerous. This implies that treating waste as a useful resource seems far-reaching. In other words, SWC and SWM are not yet improved. We have felt that the term 'sustainable waste management' has been used as an alibi rather than as a guide for strong action in Ethiopian cities. Unfortunately, its effect is already taking its toll on us. In the situation where there are heaps of waste generated from our cities, overlooking waste separation and 3R's principles implementation risks a replica of the 'Koshe' horrific disaster.

This finding implies the need for an institution specifically responsible for gluing the actors together. The Ethiopian government should not procrastinate to implement full-scale sustainable SWM.TV programmes and social media could play a role in disseminating information. The environment is a joint resource which everybody needs to care for. Particularly, since waste reduction is critical to avoid waste collection cost public education needs to be a priority.

A system of SWM which harmonises the technical requirements with the objectives of environmental protection is essential. Waste should be kept in manageable plastic waste bins that could be easily moved and comfortable for collection. This could reduce the time it takes for waste collection and service facilitation.

Given waste separation and 3R's principle's disregarded, disposal sites in the cities should be at least plotted in the master plan, demarcated, fenced, equipped with waste incinerator. They need to have leachate treatment so that they might not have negative impact on the people and environment. Moreover, after finishing their service life, the disposal sites could be used for other purposes. National project of evacuation of waste from illegal dumpsites into newly constructed sanitary landfills all over Ethiopia is strongly recommended for a better and more healthy environment. Laws to ban import of raw materials made up of plastic waste need to be implemented.

With all the challenges, however, maximising the existing opportunities could serve as a threshold to meet the aspired sustainable SWM. Further studies that could contribute to environmental awareness and local governments' commitment need to be conducted.

Author details

Abdulkerim Ahmed[1*] and Meine Pieter van Dijk[2]

1 Management Department, Wollo University, Mekelle Ethiopia, Ethiopia

2 ISS, The Hague, The Netherlands

*Address all correspondence to: aaabduky1@gmail.com

References

[1] Henry R, Yongsheng Z. Municipal solid waste management challenges in developing countries, Kenyan Case Study. Waste Management. 2005;**26**: 92-100

[2] Kaseva ME, Mbuligwe SE, Kassenga G. Recycling inorganic solid wastes: Results from a pilot study in Dar es Salaam. Resources Conservation and Recycling, Elsevier Sciences. 2005;**35**: 243-257

[3] Kassim SM, Ali M. Solid waste collection by the private sector: Households' perspective—Findings from a study in Dar es Salaam city, Tanzania. Habitat International. 2006;**30**(4):769-780

[4] Obirih-Opareh N, Post J. Quality assessment of public and private modes of solid waste collection in Accra, Ghana. Habitat International. 2002;**26**:95-112

[5] Muller M, Scheinberg A. Gender-linked livelihoods from modernising the waste management and recycling sector: A framework for analysis and decision making. Gender and the waste economy. Vietnamese and International Experiences. 2003:15-39

[6] Haylamicheal ID, Desalegne SA. A review of legal framework applicable for the management of healthcare waste and current management practices in Ethiopia. Waste Management & Research. 2012;**30**(6):607-618

[7] JICA, 2010. Preparatory Survey for Integrated Solid Waste Management in Nairobi City in the Republic of Kenya

[8] Karanja AM. Solid Waste Management in Nairobi: Actors, Institutional arrangements and contributions to sustainable development. Shaker Publishing; 2005

[9] Awortwi N. Building new competencies for government administrators and managers in an era of public sector reforms: The case of Mozambique. International Review of Administrative Sciences. 2010;**76**(4): 723-748

[10] Collins R. From Satellite to Single Market: New communication technology and European public service television. Routledge; 2005

[11] Koppenjan JF, Enserink B. Public–private partnerships in urban infrastructures: Reconciling private sector participation and sustainability. Public Administration Review. 2009; **69**(2):284-296

[12] Katusiimeh MW, Mol APJ, et al. The operations and effectiveness of public and private provision of solid waste collection services in Kampala. Habitat International. 2012;**36**(2):247-252

[13] Tilaye M, van Dijk MP. Private sector participation in solid waste collection in Addis Ababa (Ethiopia) by involving micro-enterprises. Waste Management & Research. 2014;**32**(1): 79-87

[14] Baud I, Post J. Between market and partnership: Urban solid waste management and contribution to sustainable development. GBER. 2002;**3**: 46-65

[15] Dorvil L. Private sector participation in integrated sustainable solid waste management in low-and middle income countries [doctoral dissertation]. Verlag nicht ermittelbar; 2007

[16] Van der Klundert A, Anschiitz J. The Sustainability of Alliances between Stakeholders in Waste Management. The Netherlands: ©WASTE 2001; 2000

[17] Vining J, Ebreo A. Predicting and recycling behavior from global and specific environmental attitudes and changes in recycling Opprtunities. Journal of Applied Social Psychology. 1992;**22**:1580-1607

[18] Derksen L, Gartrell J. The social context of recycling. American Sociological Review. 1993:434-442

[19] Guagnano GA, Stern PC, Dietz T. Influences on attitude-behavior relationships: A natural experiment with curbside recycling. Environment and Behavior. 1995;**27**(5):699-718

[20] Berger A. Continuous improvement and kaizen: Standardization and organizational designs. Integrated Manufacturing Systems. 1997

[21] Schahn J, Holzer E. Studies of individual environmental concern: The role of knowledge, gender, and background variables. Environment and Behavior. 1990;**22**(6):767-786

[22] Daneshvary N, Daneshvary R, Schwer RK. Solid-waste recycling behavior and support for curbside textile recycling. Environment and Behavior. 1998;**30**:144-161

[23] Hopper JR, Nielson JM. Recycling as altruistic behavior: Normative and behavioral strategies to expand participation in a community recycling program. Environment and Behavior. New Jersey: Prentice Hall; 1991;**30**:195-220

[24] Chan K. Mass communication and pro-environmental behaviour: Waste recycling in Hong Kong. Journal of Environmental Management. 1998;**52**(4):317-325

[25] Steel BS. Thinking globally and acting locally? Environmental attitudes, behaviour and activism. Journal of Environmental Management. 1996; **47**(1):27-36

[26] Selman P. Local Sustainability: Planning and Managing Ecologically Sound Places. London: Paul Chapman; 1996

[27] Van der Klundert A, Lardinois I. Community and Private (Formal and Informal) Sector Involvement in Municipal Solid Waste Management in Developing Countries. WASTE; 1995

[28] Wilson D, Whiteman A, Tormin A. Strategic planning guide for municipal solid waste management. Washington: World Bank; 2001

[29] Wilson E, McDougall FR, Willmore J. Euro-trash searching Europe for a more sustainable approach to waste management. Resources Conservation and Recycling. 2001b;**31**:327-346

[30] CSA. Ethiopian Central Statistical Agency Report. 2015. Retrieved 12 January, 2016, from http://www.csa.gov.et/images/documents/pdf_files/nationalstatisticsabstract/2006/total.pdf.

[31] Debere MK, Gelaye KA, Alamdo AG, Trifa ZM. Assessment of the health care waste generation rates and its management system in hospitals of Addis Ababa, Ethiopia, 2011. BMC Public Health. 2013;**13**(1):1-9

[32] Tadesse T, Ruijs A, et al. Household waste disposal in Mekelle city, Northern Ethiopia. Waste Management. 2008;**28** (10):2003-2012

[33] Getahun T, Mengistie E, et al. Municipal solid waste generation in growing urban areas in Africa: Current practices and relation to socioeconomic factors in Jimma, Ethiopia. Environmental Monitoring and Assessment. 2012;**184**(10):6337-6345

[34] FFE and UNEP. Assessment of the Solid Waste Management System of Bahir Dar Town and the Gaps Identified for Development of ISWM Plan. B. D. C. Administration. Bahir Dar, Forum for Environment and United Nations Environment Program. 2010. p. 14

[35] Glawe U, Visvanathan C, Alamgir M. Solid waste management in least developed Asian countries—A comparative analysis. In: International Conference on Integrated Solid Waste Management in Southeast Asian Cities. 2005. p. 5-7

[36] Glawe U, Visvanathan C, Alamgir M. Solid waste management in least developed Asian countries—A comparative analysis. In: International Conference on Integrated Solid Waste Management in Southeast Asian Cities. 2005. pp. 5-7

[37] Oduro-Kwarteng S. Private sector involvement in urban solid waste collection. Performance, capacity and regulations in five cities in Ghana. De rol van de private sector bij het ophalen van vast afval in de stad. Prestaties, capaciteit en regulering in vijf steden in Ghana. 2011. From http://hdl.handle.ne t/1765/26110.

[38] Hage O, Söderholm P, Berglund C. Norms and economic motivation in household recycling: Empirical evidence from Sweden. Resources, Conservation and Recycling. 2009;**53**:155-165

[39] Martin M, Williams I, Clark M. Social, cultural and structural influences on household waste recycling: A case study. Resources, Conservation & Recycling. 2006;**48**:357-395

[40] Williams ID, Kelly J. Greenwaste collection and the pub ic's recycling behaviour in the Borough of Wyre, England. Resources, Conservation and Recycling. 2003;**38**:139-159

[41] Addo-Yobo FN, Mansoor A. Households: Passive users or active managers? The case of solid waste management in Accra, Ghana. International Development Planning Review. 2003;**25**(4):373-389

[42] Post J, Obirih-Opareh N. Partnerships and the public interest: Assessing the performance of public-private collaboration in solid waste collection in Accra. Space and Polity. 2003;7(1):45-63

[43] Gunton H, Williams ID. Waste minimisation using behaviour change techniques: A case study for students. In: Lechner P, editor. Waste Matters: Integrating Views. Proceedings of the 2nd BOKU Waste Conference, April 16–19, Vienna. 2007. ISBN: 978-3-7089-0060-5

[44] McKenzie-Mohr D. Promoting sustainable behaviour: An introduction to community-based social marketing. Journal of Social Issues. 2000;**56**(3): 543-554

[45] Wilson DC, Velis C, Cheeseman C. Role of informal sector recycling in waste management in developing countries. Habitat international. 2006;**30**(4):797-808

[46] Van der Klundert A, Anschiitz J. Integrated Sustainable Waste Management-the Concept. Gouda; 2001

[47] Hadjieva-Zaharieva R, Dimitrova E, Buyle-Bodin F. Building waste

management in Bulgaria: Challenges and opportunities. Waste Management. 2003;**23**(8):749-761

[48] Dantata N, Touran A, Wang J. An analysis of cost and duration for deconstruction and demolition of residential buildings in Massachusetts. Resources, Conservation and Recycling. 2005;**44**(1):1-15

[49] Tam VW. On the effectiveness in implementing a waste-management-plan method in construction. Waste Management. 2008;**28**(6):1072-1080